AF567076

ANALYTICAL ATOMIC ABSORPTION SPECTROMETRY:

AN INTRODUCTION

Alfredo Sanz-Medel and Rosario Pereiro

First Edition

A volume in Coxmoor's

Spectroscopy Series of Handbooks

ANALYTICAL ATOMIC ABSORPTION SPECTROMETRY: AN INTRODUCTION

1st Edition

Authors: Alfredo Sanz-Medel and Rosario Pereiro

Series Editor: John Chalmers

13 Number ISBN: 978-1-901892-26-0
10 Number ISBN 1 901892 26 3

ANALYTICAL ATOMIC ABSORPTION SPECTROMETRY: AN INTRODUCTION

A volume in Coxmoor's *Spectroscopy Series of Handbooks*

First edition 2008

Books in *Coxmoor's* Machine & Systems Condition Monitoring Series include:
Vibration
Wear Debris Analysis
Thermography
Corrosion
Oil Analysis
Load Monitoring
Acoustic Emission & Ultrasonics
Level, Leakage & Flow
Noise & Acoustics
The Concise Encyclopedia of Condition Monitoring

Books published in association with the British Fluid Power Association include:
Principles of Hydraulic System Design
Electrohydraulics

Scientific and Technical Monographs
The Nature of the Industrial Innovation Process
Electrical and Magnetic Circuits

Conference Proceedings
Condition Monitoring 1999, 2001, 2003, 2005, 2007 (individual volumes)
COMADEM 2001
QRM 2007

Published by
Coxmoor Publishing Company
Oxford, UK
Tel: +44 (0) 1451 - 830261
Fax: +44 (0) 1451 - 870661
E-mail: mail@coxmoor.com
Printed in Great Britain by Lightning Source

CONTENTS

Chapter Three – Basic Components of Atomic Absorption Spectrometric Instruments

Chapter Four – Flame Atomic Absorption Spectrometry

PREFACE

The Coxmoor Publishing Company's ***Spectroscopy Series of Handbooks*** is intended to be a library of books introducing concisely the concepts, basic theory and methodology, together with some application examples, of the many spectroscopic techniques used today. Each volume in the *Series* will enable the reader to gain insight and understanding of the practical implementation of a particular technique, its attributes and limitations.

This exciting new *Series* is being launched this year (2008) with this volume, which is devoted to analytical *Atomic Absorption Spectrometry*, co-authored by Professor Alfredo Sanz-Medel and Dr. Rosario Pereiro. (The second book in the *Series* on nuclear magnetic resonance (NMR) of solids is in preparation. Other volumes will follow at regular intervals, each authored or co-authored by internationally recognised experts in the field.)

Atomic Absorption Spectrometry (AAS) is not a new technique, its roots can be traced back to the 19th century, and it is at the core of elemental analysis techniques; modern analytical AAS has evolved from the pioneering work in the 1950s at the CSIRO (Commonwealth Science and Industry Research Organisation) in Melbourne, Australia, by a team led by Alan Walsh. AAS measures the concentration of gas-phase atoms through their characteristic absorption of light at element-characteristic wavelengths. As described in this book, while underlying many more advanced methods of elemental analyses, AAS remains today a basic, cost-effective workhorse analytical technique for elemental analysis in many laboratories.

The chapters in this book start with an introduction to general Analytical Atomic Spectrometry (the basics, line spectra and an overview of techniques); this is followed by the theory and basic concepts in AAS. Chapter 3 describes the basic components of AAS instrumentation. Chapters 4 to 6 cover the key techniques for vaporising atoms for detection in AAS; these are flame AAS, hydride generation and cold vapour generation AAS, and electrothermal AAS, respectively. Chapter 7 covers the important area of the coupling of flow analysis techniques, including on-line chromatographic separations, with AAS detection. The book closes with a student-researcher, newcomer, user- or potential user-oriented chapter containing a Buyers Guide (including web site addresses), which lists major companies supplying AAS instrumentation, a Glossary of technical terms in AAS, a section on Standards, and a References and Bibliography section.

As the Series Editor for this new series of Coxmoor Publishing spectroscopy handbooks, I am delighted that we have launched this *Series* with this excellent book, which underlines the principles in coverage that others in the series will follow.

John Chalmers (Series Editor), May 2008.

Chapter One

AN INTRODUCTION TO ANALYTICAL ATOMIC SPECTROMETRY

1.1. BASIC INTERACTIONS OF ELECTROMAGNETIC RADIATION WITH ATOMS FOR CHEMICAL ANALYSIS.

Most molecular spectroscopy-based analytical methods available take advantage of the differential features of interacting radiation photons with the matter to be analysed. In particular, the interaction of photons with electrons of the molecules of matter is observed and measured. Photons most frequently used for such analytical purposes extend from the ultraviolet (UV: 190-390 nm) to the visible (Vis: 390-750 nm) regions of the electromagnetic spectrum and interact easily with valence electrons. Of course, the infrared (IR: 750-2500 nm) region is also much more useful in molecular analysis.

There are three main basic photon-electron interactions of the molecule that are analysed: absorption, emission and fluorescence. If the energy of a photon, $E = h.\nu$, is equal to the energy gap between electronic energy states 1 and 2 ($\Delta E = E_2 - E_1 = h.\nu$) the so-called "resonance condition" is fulfilled and the photon "disappears" as its energy is absorbed to promote the electron to the higher energy state 2 ("excited state" of the electron). Also, the electron may be promoted to an excited state by other types of energy (thermal, electrical, etc). Once excited, the electron is not stable and seeks to return to the lower energy (non-excited) level. In doing so, the energy difference $\Delta E = E_2 - E_1$ can be emitted as electromagnetic radiation of frequency $\nu = \Delta E/ h$ (h is Planck's constant: 6.63×10^{-34} J.s). When the energy used to promote the electron to the excited state is in the form of photons (photoexcitation), the spontaneous emission occurring is called fluorescence.

This tutorial book deals with analytical atomic spectroscopy. Thus, in this case the analysed matter (sample) must be in the form of atoms in the gas phase, but the main interactions we make use of for the analysis are the same: absorption, emission and fluorescence.

These interaction processes between UV-Vis photons and the outer electrons of the atoms of the sought elements can be studied and understood using the quantum mechanics theory. In the thermodynamic equilibrium between matter and interacting electromagnetic radiation, according to the radiation laws postulated by Einstein, three basic processes (quantized transitions) between the two stable energy levels 1 and 2 are possible. These three processes, which can be defined by their corresponding transition probabilities, are summarised in Figure 1.1:

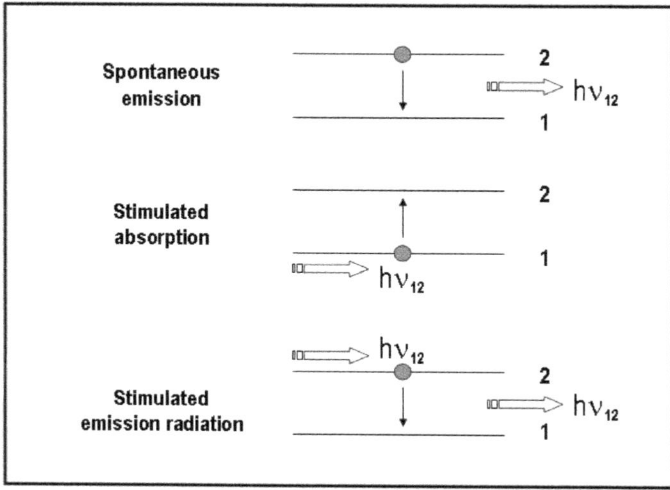

Figure 1.1. Basic interaction processes between matter and interacting electromagnetic radiation.

a) *Spontaneous emission of photons*: This process refers to a spontaneous transition of the electron from the excited state 2 to the lower energy state 1 with emission of a photon of frequency $\nu_0 = (E_2 - E_1) / h$. The probability of such a transition is usually represented by A_{21} (and corresponds to the inverse value of the lifetime of the excited state, 2).

b) *Stimulated absorption of photons*: In this case the electronic transition takes place from state 1 to state 2 in response to the action of an external radiation of the appropriate frequency. The probability of such transitions can be represented by B_{12}.

c) *Stimulated emission of photons*: This process consists of electronic transitions from the excited energy level to the lower one stimulated or in response to an external radiation of the appropriate frequency $(E_2 - E_1) / h$. The probability of these transitions is termed B_{21} (of course, a given transition only has a given probability and so $B_{21} = B_{12}$).

The first process (a) constitutes the photophysical basis of atomic emission spectrometry, while atomic absorption spectrometry is based on the second process (b). Fluorescence is, of course, the sequential combination of a stimulated absorption (b) followed by spontaneous emission (a).

The third process (c) constitutes the basis of the LASER (light amplification stimulated emission radiation) phenomenon, which is opposite to the absorption process (contradicting the common misconception that emission and absorption are really opposite phenomena). In atomic spectrometry the LASER process has been and is being used for atomisation, ionization or to build spectrochemical sources or lamps of very particular and advantageous features. In fact, the application of laser sources in this field is enormous, but the basic interaction taking place in the atomiser itself, where the sample is located for the corresponding analysis, is not a laser process (as said before the analytically useful basic interactions are absorption, emission and fluorescence).

From an analytical point of view the fact of having the analyte elements of the sample in the form of gaseous atoms is most convenient because we obtain "atomic" spectra. Such spectra (plots of the measured absorption, emission or fluorescence versus the energy of the photons employed in the experiment) are much simpler to interpret than molecular spectra.

Atomic spectra are usually a series of very narrow peaks (e.g. a few picometers bandwidth) whose analytical beauty consists of: (i), the observed peak frequency, $\nu = \Delta E/h$, will tell us the element present (because the electronic transition energy is characteristic of the element considered); (ii), at a given frequency (peak) we are looking at a given element so the measured peak-height or area will inform us about the concentration of that element in the sample.

The relative simplicity of such atomic spectra and its straightforward qualitative and quantitative information has led to the enormous practical importance and present predominance of atomic spectrometry for inorganic elemental analysis. However, it should be stressed that to gain such simplicity of spectra we have to transform the elements constituting the sample into "active" species, that is, into atoms which will interact with the photons of the UV-Vis radiation. This fact brings about at least one practical and one fundamental consequence:

- From a practical point of view we will need an "atomiser", usually a dynamic medium of high temperature where molecules of the sample are cleaved and broken down into individual gaseous atoms.

- From a fundamental point of view it should always be borne in mind that in the atomiser all molecular information is basically destroyed. The information provided by analytical atomic spectrometry is thus elemental. If molecular information is required, molecular spectroscopy should be used or we should resort to a separation technique of molecules coupled to the atomic detectors (as carried out in "speciation" analysis).

1.2. ATOMIC LINE SPECTRA AND THEIR ORIGIN.

The transitions of outer electrons of an atom may be represented as vertical lines on an "energy level" diagram, as shown in Figure 1.2 where each energy level of the outer electron possesses a given energy and is represented by a horizontal line.

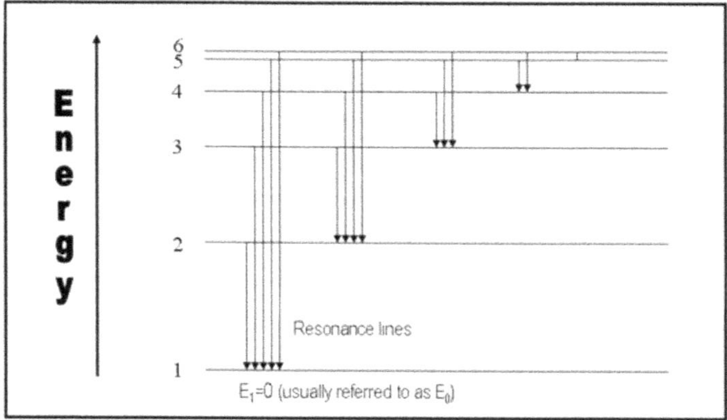

Figure 1.2. Energy levels and electronic transitions.

The vertical lines represent the energy differences between a given pair of quantized energy levels ($\Delta E = E_n - E_m$) and $E_n - E_m$ is the energy involved in such an electronic transition. Of course, each of these energy levels corresponds to an orbit of the electron around the nucleus of that atom, as illustrated by Figure 1.3.

All electronic transitions ending on the same energy level (i.e. horizontal line of Figure 1.2) are usually called "series", the most likely ones being those ending in the lowest possible energy level (the ground state) of the electron in the atom. This lower energy level is usually referred to as E_0. Of course, as depicted, there are transitions observed that have their lower energy transition level above the ground state.

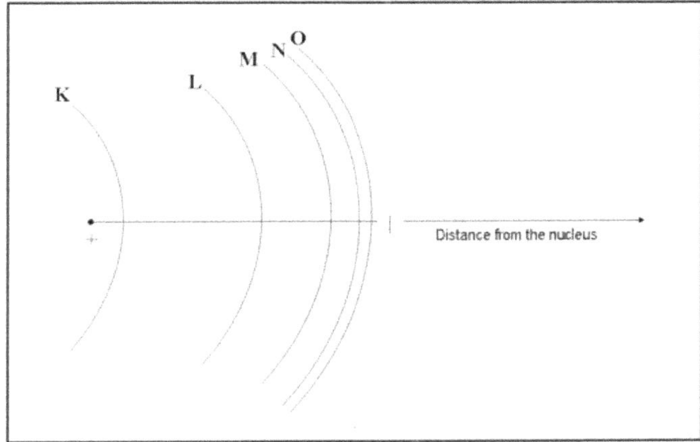

Figure 1.3. Geometric arrangement of electron shells.

Traditionally, the frequency of the light coming from such transitions (or its wavelength, $\lambda = c/\nu$, where c is the speed of light) was separated (dispersed) in a monochromator and the intensity observed for each frequency was registered on a photographic plate and constituted the "atomic spectrum". As an example, Figure 1.4 shows such photographic spectra recorded from many glowing atomic lines (or transitions) for hydrogen, sodium, helium and mercury.

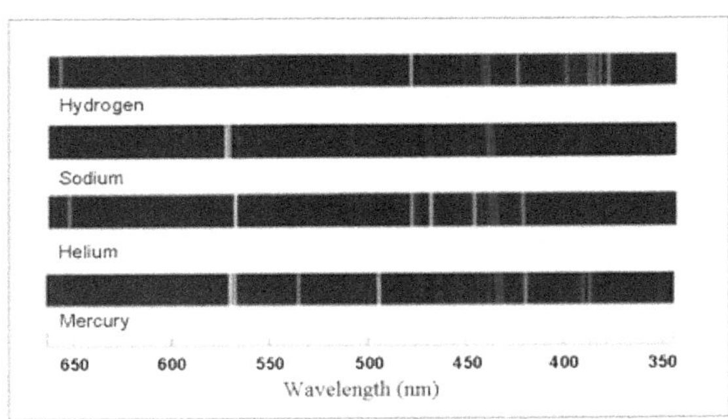

Figure 1.4. Some examples of Atomic Emission Spectra (photographic plates).

Today the detection of light is done electronically (e.g. with a photomultiplier) rather than photographically. Thus, if the observed intensity of the emitted light (radiation) is plotted versus the frequency (or wavelength) of the corresponding line (transition), a typical "atomic emission" spectrum, made of "peaks" at the wavelengths of emission, is obtained.

Thus, if a stimulated absorption of light in response to an electronic transition between a lower and a higher energy level is measured, a similar plot, per cent absorption versus frequency of the light, can be drawn, as Figure 1.5a illustrates. Such a plot represents a typical "atomic absorption" spectrum of which we will make extensive use in this book (which is specifically devoted to this type of measurement).

Finally, an "atomic fluorescence" spectrum would be the plot of the measured fluorescence intensity, coming from a cloud of atoms excited by an appropriate radiation, as a function of the frequency of the emitted radiation (Figure 1.5b).

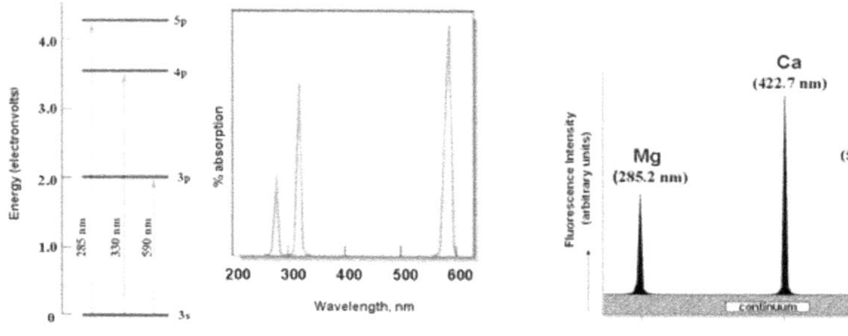

Figure 1.5. Examples of typical atomic absorption and atomic fluorescence spectra.

(a - left) The typical atomic absorption spectrum of sodium vapour and corresponding sodium atom transitions.

(b - right) A typical atomic fluorescence spectrum for the main lines of a mixture of Mg, Cd and Na atoms.

As can be seen in Figure 1.5 (atomic spectra registered with electronic detection) the atomic lines are typically represented today as individual "peaks" at the corresponding wavelength. Such lines carry very valuable analytical information: while the frequency ν ($\nu = c/\lambda$) at which the peak occurs corresponds to a given transition (characteristic of a given element), its peak area relates to the concentration of that atom (element) in the sample.

1.3. ATOMIC LINE CHARACTERISTICS.

Of course, among all possible transitions (lines of the spectrum), the selection of the most appropriate atomic line for qualitative and quantitative purposes is critical. The choice is usually the most intense line of the analyte, particularly if it is not interfered with (overlapped) by other lines arising from concomitant elements in the sample. As a rule the most intense lines are "resonance" atomic lines (i.e. when the lowest energy level in the corresponding transition is the fundamental or "ground state") level, $E_0 = 0$).

This selection of the atomic (emission, absorption or fluorescence) line characteristic of the analyte is the first step in quantitative analysis. Measurements of the intensity observed at that line are the basis of the corresponding analyte concentration determination.

As we have seen, the atomic lines in the spectrum appear as vertical lines or "peaks" (in a very narrow interval of wavelengths) due to the nature of the transition involved. That is, in molecules an electronic transition (ΔE_{el}) is usually accompanied by simultaneous changes in the molecule vibrational (ΔE_{vib}) and rotational (ΔE_{rot}) energy levels; sometimes all the three energy types may change simultaneously in an electronic transition in a molecule ($\Delta E_{mol} = \Delta E_{el} + \Delta E_{vib} + \Delta E_{rot}$). The many transition possibilities allowed, coupled with the solvent effect derived from the aggregation state of the sample in a technique such as UV-Vis spectrophotometry, in which the sample is in solution, determines that in a UV-Vis molecular absorption (or emission) spectrum the "peaks" are widely broadened. Typically the half-bandwidth of an absorption "band" in a molecular UV-Vis spectrum is around 40 nm (or 400 angstrom, Å), while in atomic "lines" the half-bandwidth observed, resulting from pure electronic transitions, ΔE_{el}, is of a few hundredths of an Å (typically 0.03-0.05 Å).

Thus, spectral interferences in atomic spectroscopy are much less likely than in molecular spectroscopy analysis. In other words, one of the strengths of atomic spectroscopy for elemental analysis is its comparatively high freedom from spectral interferences.

In fact, even atomic "lines" are not completely "monochromatic" (i.e. only one wavelength per transition). There are several phenomena that bring about certain "broadening" as well. Therefore, any atomic line shows a "profile" (distribution of intensities measured as a function of wavelength or frequency).

The theoretical study of atomic line profiles has been a classic in atomic physics; analytical atomic spectrometry has borrowed such theoretical knowledge to predict two basic aspects of the calibration function used for quantitative analysis:

a) The dependence between the intensity measured (emission, absorption or fluorescence signal) and the analytical concentration sought.
b) The dependence of intensity measured with experimental variables and parameters (a key knowledge to optimise such parameters and so achieve optimum sensitivity for a determination).

As the analytical selectivity is conditioned by the overall broadening of the lines (particularly the form of the wings of atomic lines), a basic knowledge of the broadening phenomena is necessary as well.

Therefore, it is important to summarize here the basic concepts and parameters that define in practice the atomic line "profile" as a function of wavelength (or frequency). From a practical perspective Figure 1.6 illustrates a typical plot of such a "profile" observed for an atomic (it could be absorption, emission or fluorescence) line. As shown in the Figure, the main parameters defining an atomic line are:

I_m: Intensity of the emission, absorption or fluorescence measured at the line maximum.

λ_m: Wavelength measured at the line maximum.

$\Delta\lambda$: Half bandwidth at half height ($I_m/2$) in wavelength units.

I_b: Intensity measured for the background at λ_m.

According to Figure 1.6 the total analytical signal, I_{net} (of the profile) would be given by:

$$I_{net} = \int_{-\infty}^{+\infty} [I_m(\lambda) - I_B(\lambda)] d\lambda$$

(i.e., the area below the atomic line with corrected background at every point).

At the maximum of the line (peak height): $I_{net} = I_m - I_b$.

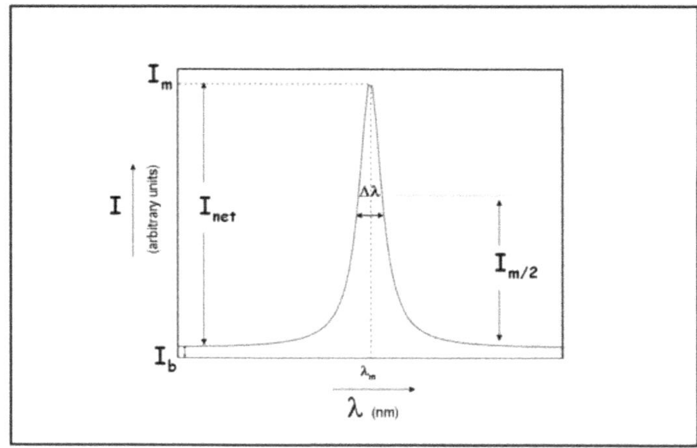

Figure 1.6. A typical atomic line "profile".

Figure 1.6 illustrates a typical distribution of intensities around the λ_m, showing the broadening of the line around that value. In the following section we will consider in more detail the main broadening phenomena that bring about the characteristic profiles of atomic lines.

1.4. ATOMIC LINES SPECTRAL WIDTH.

Let us illustrate the concept of broadening and final spectral bandwidth of atomic lines, using an "absorption" line profile (the subject of the book) for the sake of simplicity. In fact, the concept and theoretical treatment given below should be similar when describing the profiles and spectral bandwidths of emission (or fluorescence) atomic lines.

As stated before, an atomic absorption line is the result of a "pure" electronic transition from a lower to a higher energy state of the atom stimulated by a photon of the appropriate frequency. Notwithstanding what was stated for Figure 1.2, these absorptions are not infinitely fine vertical lines at frequency $\nu_0 = \Delta E/h$. The lines exhibit a certain finite width around the maximum absorption wavelength. That is, atomic lines are also broadened and the extent of such broadening can be measured as the "half-width at half-height" of the line, $\Delta\lambda$. This parameter means the width of the profile, % absorption versus wavelength, at the point at which the maximum absorption is halved.

In order to describe broadening of atomic lines (whatever the wavelength considered), in the profile it is preferred in atomic absorption spectrometry to express the "half-width" in terms of frequency, $\Delta\nu$ (cm^{-1} or s^{-1}).

Thus, as $\nu(s^{-1}) = c/\lambda$, which after differentiation gives $\Delta\nu(s^{-1}) = (c/\lambda^2)\,\Delta\lambda$

[and $\Delta\nu(cm^{-1}) = \Delta\lambda/\lambda^2$].

Although there are other possible causes for broadening of atomic lines (e.g. interactions of the atoms with electrically charged particles or, at higher concentrations, with each another), the profile of an atomic line is governed mainly by the combined effect of the following processes: natural, Doppler and Lorentz broadening.

1.4.1. The natural broadening of lines.

This type of broadening is associated with the degree to which the energy of the levels is defined. By Heisenberg's Uncertainty Principle, from the point of view of quantum electrodynamics, the energy, E, of a system in a given energy state and the time, t, during which that system remains in such state cannot be known accurately simultaneously. This is formulated as:

$$\Delta E \cdot \Delta t \geq h/4\pi$$

In our case this means that if we consider a given atom, we must have uncertainty in the time spent by the electron in the excited state before spontaneous deactivation.

In other words, the uncertainty in the finite lifetime, τ, of the levels between which a transition takes place determines an actual broadening of the energy levels:

$\Delta E \geq (h/4\pi)/\Delta t$

As the ground level is stable ($\tau = \infty$), it is the width of the upper level that is the only broadening of significance for resonance transitions (see Figure 1.7). Thus, the natural width of a resonance line can be defined as:

$\Delta\nu_N = B_{12}/2\pi = 1/2\pi\tau$

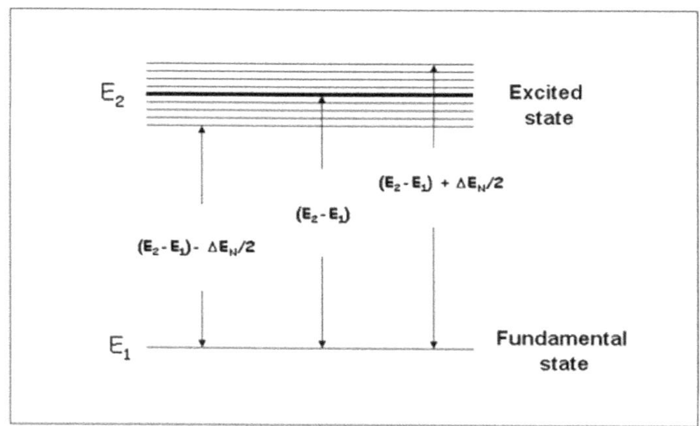

Figure 1.7. Natural line-broadening origins.

Different spectral lines have different natural widths (e.g. Hg at 253.7 nm, $\Delta\nu_N$ = 5.3 x 10^{-5} cm^{-1}; Cd at 228.8 nm, $\Delta\nu_N$ = 2.7 x 10^{-3} cm^{-1}) but it is worth noting that in most cases the natural width of atomic lines does not exceed 10^{-3} cm^{-1} and so, as we will see later on, this broadening can then be ignored against other broadening phenomena.

1.4.2. Doppler broadening.

The Doppler broadening in atomic lines arises from the random thermal motion of atoms, in the high temperature atomiser, relative to the observer or detector. Similar to the Doppler effect distorting the frequency of a sound emitted by a moving object (see Figure 1.8), if an electronic transition stimulated by the absorption of radiation of frequency ν_0, takes place in an atom moving at a speed V_x in the line of the observer, the measured frequency of absorption by the atom is displaced by $\Delta\nu_D$, where

$$\Delta\nu_D = \nu_0 \cdot (V_x/c)$$

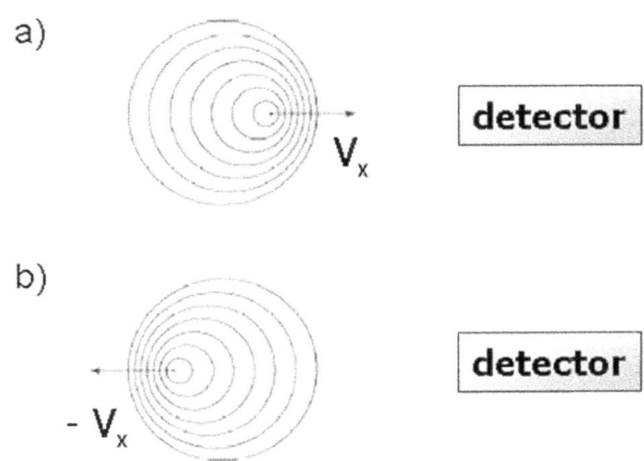

Figure 1.8. Schematic representation of the Doppler effect. As a consequence of the velocity V_x of the atom in the line of sight, the observed frequency of absorption (ν) by the atom is displaced by $\Delta\nu$. When the atom moves in the opposite direction to the detector, situation b), ν is higher than ν_0, while the value of ν is less than ν_0 for situation a).

If the atomic vapour in the atomiser (e.g. a flame) is in thermodynamic equilibrium, it can be demonstrated that the motion of atoms there can be described by the Maxwell distribution and then the Doppler half-width of the line is equal to:

$$\Delta\nu_D = (2\,\nu_0 / c)\,(2 \ln RT/W_{at})^{1/2}$$

and, if numerical values are given to the constants,

$$\Delta\nu_D = 7.16 \times 10^{-7}.\ \nu_0\,(T/W_{at})^{1/2}$$

(T the absolute temperature in K, W_{at} is the atomic weight of the moving atom and R is the gas constant).

It is important to note that for a given analyte Doppler broadening is independent of the concentration of the absorbing atoms in the atomiser cell and is proportional to the square root of the absolute temperature.

Doppler broadening exhibits a Gaussian profile and its value is usually two orders of magnitude higher than natural broadening. Table 1.1 illustrates the calculated values of $\Delta\nu_D$ for atomic lines of three chemical elements (atoms) at two temperatures. The Doppler effect is the predominant effect in the uppermost part of the atomic lines observed in a flame and describes quite accurately the observed profiles above the half maximum position of the lines. Also, this is the main factor determining the experimental width observed in the lines emitted by the hollow cathode lamps used for atomic absorption.

Atomic line (Å)	Atomic weight (amu)	$\Delta\nu_D$ (cm^{-1}) (500 K)	$\Delta\nu_D$ (cm^{-1}) (3000 K)
Cs 8521	132.91	1.6×10^{-2}	4×10^{-2}
Li 6708	6.94	9.1×10^{-2}	2.2×10^{-1}
B 2498	10.81	2×10^{-1}	4.8×10^{-1}

Table 1.1. Calculated values of Doppler broadening ($\nu\Delta D$) of some atomic lines used for chemical analysis of Cs, Li and B at two temperatures.

Note: It is important to stress that even at a relatively low temperature (500 K) Doppler broadening, $\nu\Delta D$, is much greater than $\nu\Delta N$

1.4.3. Lorentz broadening.

This effect, together with the Doppler effect, make the predominant contributions to the eventual shape, width and position of an absorbing line.

The Lorentz broadening, $\Delta\nu_L$, is observed when the pressure of a foreign gas in the atomiser is increased. Experience shows that collisions with the gas cause broadening, asymmetry of the line profile and a displacement of the maximum of the line, relative to its initial position. Both, displacement of the original λ_{max} (maximum shift) and line asymmetry can be easily evaluated in a quantitative manner (see Figure 1.9). It is known that different foreign gases have different effects on the broadening and shifting of the lines. Also, such changes, for a given gas, are proportional to the change in total pressure in the atomiser.

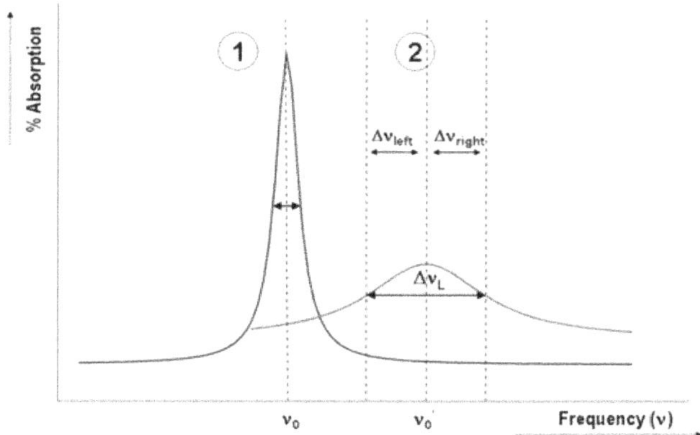

Figure 1.9. Effects of the pressure (Lorentz broadening) on line profiles: 1, Normal profile ; 2, Lorentz broadening (broadened and shifted profile).

Different theoretical approaches have been proposed to interpret this Lorentz or "pressure broadening". While the "collisional" theory offers the best fit equations to explain broadening at the centre of the line (which is the most important part in atomic absorption, as we will see in the next Chapter), the so-called "statistical" theory describes better these effects at the wings of the line.

According to the work of Lorentz, back in 1905, atomic emission of light comes from the harmonic vibration of electrons within the atom. In the presence of a foreign gas the atoms may collide with the foreign particle and so vibration is interrupted in the collision, while vibration is renewed at the same frequency immediately afterwards.

In the light of this collisional theory it has been shown that the Lorentz halfwidth depends on:

$$\Delta\nu_L = 2.6 \times 6.02 \times 10^{23} \cdot \delta^2 \cdot P\,[2/(\pi RT) \cdot (1/W_{at} + 1/W_{mol})]^{1/2}$$

where P is the pressure of the gas, W_{at} the atomic weight of the atom, W_{mol} the molecular weight of the gas and δ^2 is the effective cross-section for the atom-molecule collision considered.

From a quantum mechanics point of view, the collision would modify the excitation state of the atom, altering in that way the lifetime of the excited level. As a result, a broadening of the line qualitatively similar to $\Delta\nu_N$ (but much more intense) would take place.

It should be stressed at this point that the experimental absorption profiles of analyte atoms in a flame (the most common atomic absorption atomiser) are governed by both Doppler and Lorentz broadening: the centre of the line primarily by the Doppler effect and the wings by the Lorentz effect (see Figure 1.10).

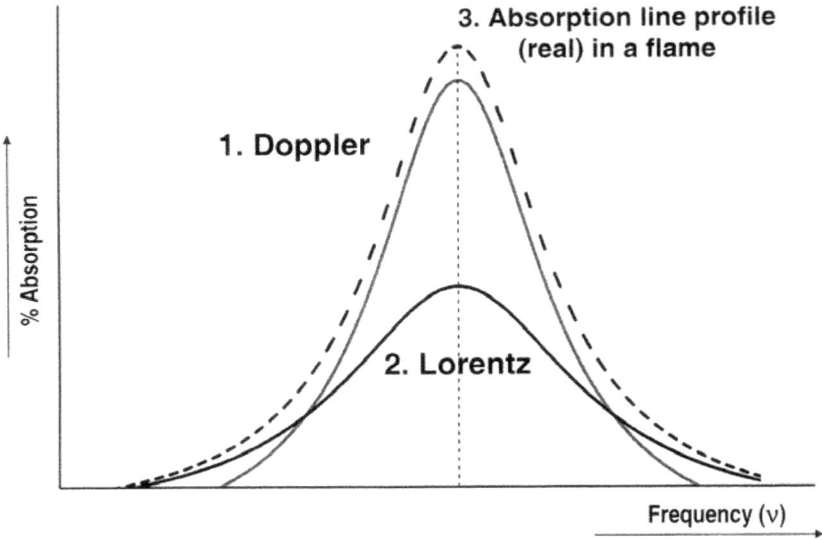

Figure 1.10. A typical symmetrical "profile" of an absorption line in a flame: the centre of the line behaves mostly as if Doppler broadened, while the wings are mostly Lorentz – broadened.

1.4.4. Self-absorption effects.

In emission measurements, it is well known that at low atom density levels, increasing the concentration of the analyte in the atomiser increases proportionally the height of the observed emission line. However, at a particular concentration, C, the profile of the atomic line "broadens" rather than increasing its height (see Figure 1.11). This concentration effect is due to self-absorption of emitted radiation, by the excess of analyte atoms in the ground level state. Of course, as the maximum absorption occurs always in the centre of the line, this effect is much stronger in the centre than in the wings of the line. Moreover, the phenomenon will affect particularly resonant lines (where the lower energy level of the transition is the ground level).

Self-absorption effects, although better described for emission measurements, are also present as a cause of broadening in all atomic lines. For instance, in a flame, where all the analyte atoms are roughly at the same temperature, this self-broadening of the lines is mainly responsible for the loss of linearity of calibration curves (absorbance versus concentration) at higher concentrations.

To conclude this section it should be mentioned that when a temperature gradient exists in the atomiser and the absorbing analyte atoms are at a lower temperature than the emitting ones (e.g. in an electric arc discharge) the profile of the absorption is narrower than that of the emission. Thus, strong self-absorption may take place just around λ_{max} and it may be so intense there that the observed atomic line splits into two distinct lines (see Figure 1.11). This is the "self-reversal" phenomenon, often observed in resonance emission lines in lamps used as spectral light sources at high atomic density emissions, as we will see in Chapter 3 when describing the Smith-Hieftje "background corrector".

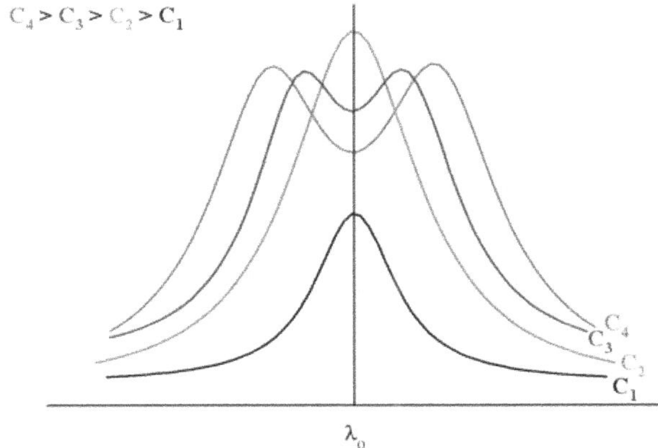

Figure 1.11. Autoabsorption and self-reversal of a spectral line. C indicates analyte concentration.

1.4.5. Other broadening processes.

Other broadening effects, affecting the profile of atomic lines, have been observed in different atomisers and spectral sources and under different conditions. Although none of these broadening processes are of general practical importance in atomic absorption spectrometry using a flame, for the sake of completeness we will mention the following effects:

- Stark broadening which takes place in the presence of (high) electric fields.

- Zeeman broadening occurs when working in a magnetic field. Of course, in flames these processes are of no significance but Zeeman broadening should be considered also for "background correction" (see Chapter 3) in electrothermal atomic absorption techniques (as described in Chapter 6).

- Holtsmark broadening is related with Lorentz broadening. Such a process is due to collisions between atoms of the same kind (e.g. the analyte) and thus it is also called "resonance" broadening in classical books. Of course, as relative concentrations of the analyte are so low, resonance broadening is of negligible importance compared with Lorentz broadening.

Finally, "hyperfine structure" should be mentioned as it accounts for an apparent broadening of lines, which is in fact due to overlapping of individual lines (because the resolving power of the monochromator used is not high enough). As it is known, interactions between the spin of the nucleus of the atom (different for different isotopes) and the resultant spin moment of its electrons may result in a splitting of the corresponding energy levels, that is in a given "multiplicity" of expected energy levels in that atom (e.g. Hg vapour emission).

An interesting example of isotopic "hyperfine structure" in atomic spectrometry is the emission of a Hg lamp: the Hg 253.7 nm line is in fact an emission of multiple peaks resulting from ten possible excited energy levels (derived from even-numbered Hg isotopes).

1.5. A COMPARATIVE OVERVIEW OF ANALYTICAL ATOMIC SPECTROMETRIC TECHNIQUES.

From the preceding narrative it follows that analytical atomic spectrometry is not just one technique. At least, atomic absorption, atomic emission and, much less used, atomic fluorescence techniques are available.

Usually a dissolved sample is used for analysis to form a liquid spray of the sample that is delivered to the atomiser (e.g. a flame). However, direct solid analysis is also possible (and is today used throughout the world) by using special atomiser/excitation sources such as arcs, sparks, lasers or glow discharges.

From a practical point of view, routine inorganic elemental analysis is mainly carried out nowadays by analytical atomic spectrometry techniques based on the measurement of "photon" interactions (e.g. interactions between photons of electromagnetic radiation and the matter under scrutiny). Current "workhorses" for such analyses include both dissolved sample and direct solid analysis techniques.

1.5.1. Dissolved sample analysis techniques.

Flame-atomic absorption spectrometry (FAAS), electrothermal atomisation atomic absorption spectrometry (ETAAS) and inductively coupled plasma-optical emission spectrometry (ICP-OES) are probably today's "workhorses" for routine analysis of dissolved samples.

Instrumental development, in addition to analytical applications, which increased after commercialisation of AAS in the 1960s and of ICP-OES in the 1980s, has been profound and extensive since. Knowledge about such techniques is now well-advanced and so new spectacular breakthroughs are not likely. Thus, the capabilities and analytical limitations are well known for comparing their respective analytical performance characteristics. An attempt in that direction is summarized in Table 1.2 which gives a general comparative assessment, based on many years of evaluating the main "pros" and "cons" of each of these three popular and routine AA techniques.

	Flame-AAS	Electrothermal-AAS	ICP-OES
General Advantages	Simple and reliable Most widespread Moderate interferences Ideal for unielemental monitoring in small labs High sample throughput	Sub-ppb (μg . L^{-1}) DLs Microsamples (< 1 ml)	Multielemental High temperature (7000 °K) Low matrix interferences High dynamic range
Cost	Low cost	Higher cost	High instrument cost
Limitations	Uni-elemental Sub ppm (mg . L^{-1}) DLs Low temperature (refractary compounds problem) Only metal-metalloids (> 60 elements)	Unielemental Not so easy to use Carbide formation For metal-metalloids (50 elements)	Serious spectral interferences Sub ppm-ppb (μg . L^{-1}) DLs Expensive to run Also for some non-metals (> 70 elements)

Table 1.2.- Comparative advantages and limitations of the most common atomic "workhorses" of dissolved samples analysis

Taking the risk inherent in excessive generalizations, perhaps we could say that:

(i) FAAS dominates elemental inorganic analysis carried out in rather small laboratories and when only a few analytes (probably at mg.L^{-1}, ppm, levels) have to be determined.

(ii) When ppb level (μg.L^{-1}) sensitivity is required the technique of choice is

ETAAS, at a cost of simplicity, robustness and speed of the analysis as compared to FAAS.

ICP-OES appears to have become the most popular routine technique for inorganic multielemental analysis of dissolved samples, even if the initial investment and subsequent running expenses are much higher than those needed for AAS.

Of course, flow injection analysis (FIA) strategies have become commonplace in most laboratories opening new opportunities to sample handling, pretreatments and manipulations for atomic spectrometry methods.

AAS instruments, flame and electrothermal atomisation based, are today widespread all over the world, both in developed and developing countries. They are less complex systems than multielement techniques (e.g. ICP-OES) and so instrumentation for single-element AAS is considerably less costly. Thus, this book is devoted to the AAS technique.

1.5.2. Direct solid analysis techniques.

The rapid growth of ICP-OES might obscure the fact that the market size for direct solid analysis is still close to that for dissolved samples. Spark source-optical emission spectrometry (SS-OES) and X-ray fluorescence spectrometry (XRF) are very well established routine techniques, which play a very important role in industrial analytical chemistry of solid materials. Their importance today to monitor industrial processes, metallic and non-metallic raw materials and final products, cannot be overemphasized. They have no competitors in that application. Both techniques are so well implemented in routine laboratories that it is sometimes said that there is nothing new in X-ray or spark source based analyses.

More recently, other spectrochemical sources such as lasers, and also glow discharges with optical emission spectroscopy measurements, are gaining acceptance for the direct analysis of metallic, non-metallic and semiconductor solid samples.

While SS-OES is clearly an atomic technique, where atoms are formed directly from the solid sample by virtue of the high energy of an electrical spark, the inclusion of XRF among the atomic techniques deserves some explanation. In this latter technique the composition of a solid sample is interrogated at room temperature and so no atoms are formed during the analysis. In XRF a primary beam of X-Rays is used to excite and eject electrons from the inner shells of the atoms (e.g. shells K or L) of a solid. After such excitation, electrons of the outer shells fall back spontaneously to fill in the "holes" vacated and the energy difference for such inner electrons transitions is given back as electromagnetic radiation (fluorescent or secondary X-Rays).

The information given by a X-Ray fluorescent spectrum relates basically to the elements in the sample (not to the molecules). In other words, although strictly speaking XRF is a technique where the analysed sample is not transformed into atoms, the information provided is mainly atomic (i.e. of the individual elements present in the solid sample).

ICP-MS and ICP-OES (versus AAS)

- Multielemental character (simultaneous multielement analysis possible).
- High speed of analysis.
- Semiquantitative rapid analysis (easier by ICP-MS).
- Continuous operation.

ICP-MS exclusive features

- Elemental detection in the ng.L-1 range (or below).
- Simple mass spectra.
- Isotope measurement capability allowing for:
    - Confirmation of the presence of multisotopic elements.
    - Isotope dilution analysis.
    - Stable isotope tracers applications (e.g. to follow the metabolism of elements).

Table 1.3.- ICP-MS and ICP-OES main relative analytical merits.

Finally, to complete the picture of current atomic spectrometric techniques for inorganic elemental analysis, due reference to the increasing analytical importance of inductively coupled plasma-mass spectrometry (ICP-MS) should be made here. As suggested by the name, ICP-MS is the combination of an ICP with a mass spectrometer. During ICP-OES development it became clear that an argon ICP was also a most efficient ion source to generate singly charged ions from the elements within a sample. It happened also that the kinetic energies of the ions formed in the argon plasma were between 2-10 electronvolts, an energy interval most appropriate to be resolved by an inexpensive quadrupole mass analyser. Given this situation it is not surprising that coupling an argon ICP to the ubiquitous mass spectrometer technique was a tremendous success. The analytical performance characteristics of such a synergic combination, ICP-MS, are usually striking. As Table 1.3 shows, in a comparative manner, ICP-MS performance is clearly superior to its "brother" multielemental technique, ICP-OES. Apart from the extreme sensitivity of ICP-MS and its relative spectra simplicity, probably its most differential (advantageous) feature versus photon-based measurement techniques is the ability of ICP-MS to determine the presence of isotopes and to measure readily isotopic ratios. This capability gives to ICP-MS exclusive applications, including the flourishing of isotope dilution-ICP-MS methods for:

a) Accurate determinations of trace and ultratrace levels in a wide variety of samples.

b) "Stable isotopes tracers" use in living organisms in order to follow the metabolism of essential and toxic elements. Traditionally, radioactive tracers were employed for this purpose and most of the present knowledge about metal metabolism derives from experiments based on the administration of the corresponding radioactive isotopes to the studied living organism. Their eventual transport and distribution in the different organs is usually followed by radiochemical measurements. Of course, application of this radiochemical technique in humans has always been hindered by the detrimental effects of γ-radiation to the body. With the advent of ICP-MS a parallel strategy using non-ionizing radioactive isotopes seems straightforward.

To conclude this section, however, it must be pointed out that ICP-MS is still a comparatively expensive analytical tool with both a high price and maintenance costs. Moreover, its effective operation so far demands highly qualified personnel. Both factors hinder its widespread use in many routine laboratories.

1.6. SCOPE OF THE BOOK

Around fifty years have elapsed since the successful introduction of atomic absorption spectrometry by Alan Walsh (of the Australian Commonwealth Scientific and Industrial Research Organisation, CSIRO) for elemental analysis. At this moment in time it is comparatively rare to find research publications on new AAS fundamentals or even new AAS applications. Notwithstanding, analytical AAS has become an elemental analysis routine tool that is most firmly established and applied worldwide. What is more, it can be safely said that in today's routine laboratories AAS is the most popular, inexpensive and easy-to-use technique among all the techniques available for elemental analysis (see earlier section).

For these reasons we have restricted the scope of this tutorial book to the study of the popular, most efficient though, AAS technique trying to provide an up-to-date account of its fundamentals, instrumentation, special techniques and applications. To do so, the atomic absorption experiment and photophysical law governing such processes are reviewed. Then, the main components or units that when appropriately assembled, constitute an AAS instrument are described in detail to set the foundations of modern spectrometers for AAS measurements.

Once the basic instrumentation for AAS measurements have been covered, we have selected to write on those AAS-based techniques that are now well established and of recognised value to providing robust and fit-for-purpose analytical results. Rather than following the usual trend in several books of writing a bulky part of detailed applications to analyse particular samples or to determine particular elements, we preferred the approach of studying progressively more complex AAS techniques, which have been introduced to solve problems of lack of sensitivity, selectivity or easy sample handling in elemental determinations, as carried out worldwide in routine laboratories. Of course, at the end of each of these chapters a section of selected applications illustrating the relative merits and strength of each particular AAS technique is included.

Thus, the techniques using a classical flame as atomiser are considered first (chapter 4), while the following chapter deals with flame sensitivity problems and how to overcome them by resorting to analyte volatile species generation at room temperature from the dissolved sample, (of course, the usefulness of hydride or cold vapour generation techniques is not restricted to AAS, but it is without question that such higher sensitivity techniques, so popular in atomic spectrometry these days, were born and developed mainly using absorption measurements).

Unfortunately flame atomisation with volatile species generation is restricted to a few analytes (able to form such volatile species by adequate reactions in the aqueous phase). Thus, electrothermal (ET) atomisation is then reviewed (Chapter 6) and described as an advantageous alternative to the flame. Its greater sensitivity coming from a confined and better controlled atomisation of the whole sample (not just the analyte) explains the ETAAS superior performance for ultratrace determinations ($\mu g.L^{-1}$ concentrations of the analyte in the dissolved sample).

The following chapter is devoted to flow injection analysis (FIA)-atomic absorption spectrometry. Although FIA was born originally as a practical technique for automation of serial assays, it has evolved well beyond that original point. Today FIA offers a most convenient and straightforward approach to automate, and also to enhance, all preliminary steps for atomic spectrometric detectors in general and for atomic absorption techniques in particular. Also, the on-line coupling of chromatographic separations to AAS has demonstrated great potential, in particular for metal speciation purposes. We selected this topic as the concluding chapter due the high degree of compatibility between AAS and flow techniques but, above all, for the practical importance of such couplings to solve real problems.

A Buyer's Guide and Reference Section completes this book.

Chapter Two

THEORY AND BASIC CONCEPTS IN ATOMIC ABSORPTION SPECTROMETRY

2.1. GENERAL INTRODUCTION

We should start this chapter by stressing again that AAS continues to be the most popular and utilised atomic spectrometry technique for routine elemental determinations after more than 40 years since commercial instrumentation became available. This position confirms the practical strengths of AAS when compared with other atomic techniques (see Table 1.2 in Chapter 1) with which it is competitive for such elemental analysis. Of course, after such a long period of evolution, the theory and basic concepts of "analytical" AAS (as conceived and developed by Alan Walsh **[Walsh (1991)]**) are very well established. Thus, it is not surprising that there has been clear evidence of a slow down in research into the basics of AAS, on novel methods and even on new applications as well. The "turning point" in AAS publications seemed to occur at the end of the 1980s, a fact taken as the basis for a controversial and pessimistic prediction of the decline and "death" of AAS by the year 2000 made in 1989 **[Hieftje (1989)]**.

At the beginning of this new millennium, however, we may say that atomic absorption-based techniques are still "alive and kicking", particularly considering the high number of commercial AAS instruments used and still sold worldwide. What is more, novel analytical AAS concepts and developments are still being investigated in some research laboratories **[Bolshakov *et al.* (2006), Hywel Evans *et al.* (2005), Hywel Evans *et al.* (2006)]**. Among these research efforts, an attempt to transform AAS into a viable and competitive multi-element method of analysis deserves special mention. This encompasses the use of continuum sources in combination with high spectral resolution systems (e.g. "echelle" spectrometers and charge coupled devices (CCDs)), being now well established they are the basis of new commercial instrumentation available in today's market.

A second new concept and development (not yet commercialized) derives from the use of diode lasers, instead of classical hollow cathode lamps, to reduce the size and cost of future AAS instruments **[Zybin et al (2005)]**. Another area of basic research is the work on new concepts and instruments arising from the use of conventional lasers in connection with cavities (as atomisers) to enhance the absorption and so the sensitivity of AAS determinations **[Emig (2002)]**.

Finally, studies on introduction of volatilised analytes into the atomiser, on graphite furnaces or glow discharges as alternative atomisers, are still appearing in the literature, while renewed efforts are still being devoted to modernization, automation (e.g. with flow techniques) and, also, to miniaturisation of conventional AAS instruments in several laboratories around the world.

In spite of these on-going research activities, it is fair to say that the real present strength of AAS, its proven reliability, robustness and efficiency for real-life elemental analysis, derives mainly from the original Alan Walsh invention (his first patent in AAS occurred in 1953) where the concept of using a hollow cathode lamp as the light source for absorption measurements was decisive to the success of the technique. This exceptional light source was proposed originally in combination with a flame to atomise the sample to be analysed. Today it is clear that the nature of that light source and of the atomiser employed will determine the actual strengths and shortcomings of the AAS analytical methods based on such instrumentation.

To understand why the hollow cathode lamp performance is central to the success of AAS elemental determinations, why the type of atomiser used will determine dramatically the detection limits achieved or why the necessary straight-line geometry of AAS optics make this technique usable for only one (or just a few) element(s) at a time (perhaps its main drawback when compared with ICP-OES) are of utmost importance for an effective and efficient use of AAS in the laboratory. To help to understand these practical issues, which are of critical importance to this book, the fundamental concepts of AAS measurements and its inherent analytical strengths will be considered in the following sections.

2.2. THE BASIC AAS EXPERIMENT.

As explained in the first chapter, atomic absorption is the process that occurs when a ground state electron of the desired atom (analyte) absorbs energy, in the form of light of a given frequency, from an external light source and so the electron is excited to an upper energy level state of the atom.

The basic AAS experiment is easily illustrated by the simple schematic shown in Figure 2.1a. Such a self-explanatory schematic, common to all absorption spectroscopic techniques, shows that the main components needed to carry out AAS measurements are:

(a) the light source (which provides the external energy).

(b) the atom cell (where the sample is atomised).

(c) the monochromator (where the appropriate frequency is selected).

(d) the "light-read-out" system, integrated by a transducer or detector (transforming the light into a measurable signal, e.g. an electric current) and the electronic read-out system of such a measured signal.

With such a system, appropriately assembled, (see Figure 2.1a) the amount of light energy absorbed by the analyte atoms at the selected wavelength can be easily measured by doing two consecutive measurements: (i) the incident light intensity or "blank" (I_0), without sample in the atom cell, is measured first; (ii) then the light intensity exiting the atom cell (I), after the sample (really the analyte atoms in the light path) has absorbed the light.

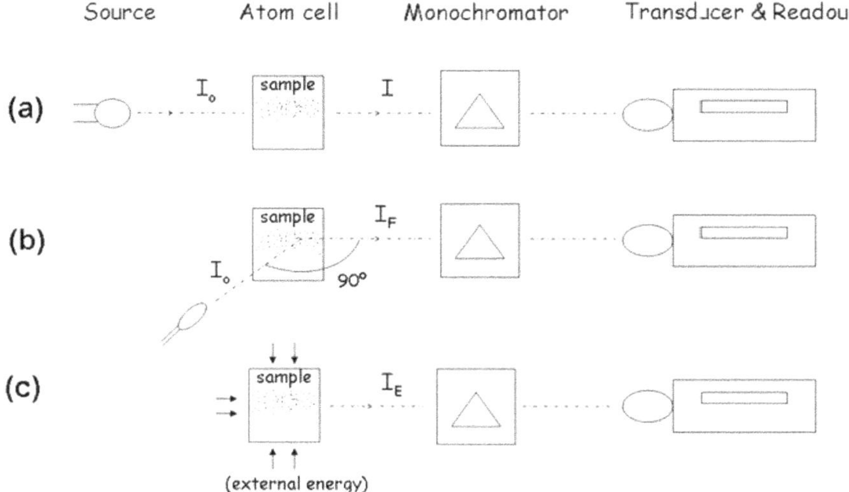

Figure 2.1. Schematics of the basic AAS experiment (a) and the main differences with the other main atomic spectrometry techniques of fluorescence (b) and emission (c).

In the absence of spurious phenomena (interferences) the amount of light energy absorbed by the analyte present in the sample will be $I_0 - I$. This amount, $I_0 - I$, will increase as the number of analyte atoms in the light path increases. As we will see later, there is a relationship between the measured absorption and the analyte concentration in the original sample. Once the exact relationship between these two measurements is clearly established (e.g. using known analyte concentrations of appropriate standards), that is, once the "calibration curve" is defined, analyte concentrations in unknown samples can be determined by measuring the amount of light they absorb. Figures 2.1b and c illustrate schematically, compared with the AAS experiment (Figure 2.1a), the differences between the three main atomic spectrometry techniques: Figure 2.1b shows that in atomic fluorescence spectrometry (AFS) the basic instrument components are the same as with AAS, only the geometry of the AAS arrangement changes as the external light source (used for photoexcitation of the analyte) has been rotated 90º with respect to the straight-line optical axis used in absorption measurements. The AFS measurement consists then of measuring the fluorescence emitted at a given solid angle (I_F) by the excited atoms.

Finally, as shown by Figure 2.1c, the basic atomic emission experiment is conceptually the simplest one, because it does not need an external light source; the excited sample in the atomiser acts as such. Here, the basic components, except the source, are again the same. However, now, some type of energy device (provided by a spark, a flame, a plasma, etc.) is needed to excite the sample for atomic emission spectrometry (AES, also known as OES from "optical emission spectrometry"). Then, the spontaneous emission radiation from excited analyte atoms occurs and the amount of light emitted by them (I_E) at the selected frequency may be measured.

Of course, as for the absorption process, both I_F and I_E can be related to the analyte concentration in the sample and so the corresponding "calibration curves" can be established. Thus, just as mentioned for AAS measurements, both alternative experimental arrangements (b and c) can enable the determination of analyte concentrations in unknown samples. They constitute the basis of the other two main atomic measurement arrangements, namely atomic fluorescence spectrometry and atomic emission spectrometry, respectively.

2.3. THE ABSORPTION COEFFICIENT CONCEPT.

There are different methods of measuring "absorption" in an atomic absorption process, depending on the experimental magnitude to be measured:

a) The "integrated absorption", where the area under the atomic absorption line profile (Figure 1.6) is measured.

b) "The total energy absorbed" from a continuum source, emitting a continuum spectrum, at the absorption line.

c) The "line absorption" method, where we measure the relative absorption of light, at the absorption wavelength, from a source that emits a line spectrum.

The latter method is the most practical one and it is called, after its inventor, the Walsh method, and the measurement constitutes the basis of routine AAS measurements today for analytical determinations. In such "analytical" AAS measurements, a line source (usually a hollow cathode lamp) is used as the critical component of the AAS experiment. As indicated in the preceding chapter, atomic absorption spectra, as opposed to the classical absorption bands observed in conventional molecular absorption spectrophotometry, show up in the form of several individual narrow lines. The frequency (wavelength) of such lines characterises the particular absorbing atoms (analyte).

Thus, using a hollow cathode lamp (HCL) as the source, an appropriate emission line of wavelength maximum λ_0 and intensity I_0 is selected with a monochromator.

This light is then passed through the atomic cloud of free atoms of the analyte (in the flame), contained in an absorption volume of path-length L. The atomic vapour in the flame absorbs the radiation emitted by the hollow cathode lamp virtually at the maximum of the element's absorption line, λ_0. As shown in Figure 2.1a, the intensity emerging from the atomiser, I (that part of I_0 which was not absorbed), is called the "transmitted" light and follows the general law of absorption, the so-called **Beer's law**:

$$\mathbf{I = I_0 e^{-k_\upsilon L}}$$

where k_υ is defined as the "absorption coefficient" at the wavelength of measurement of the atomic vapour being analysed and L the optical path in the atomiser cell (e.g., flame). The magnitude of k_υ characterizes an absorption line as strong to weak and, for a given element, will depend on the frequency of the selected line (analytical line) and on the concentration of the analyte absorbing atoms. For general instrumental quantitative analysis we usually measure a property of the desired atoms (or molecules), which varies linearly with the concentration of such atoms (species) of interest in order to establish a typical "calibration curve".

As Beer's law shows, I depends exponentially on k_υ. For this reason it is commonplace to use a more convenient form of this formula showing a linear relationship between the measured property and the desired values of analyte concentration. Such a property is the Absorbance, Abs, defined as:

$$\mathbf{Abs = -\log (I/I_0) = \log I_0/I}$$

Thus, if this concept of Abs is introduced in the previous Beer's law equation we get:

$$\mathbf{Abs = \log I_0/I = \log e^{k_\upsilon L} = 0.4343\, k_\upsilon L} \qquad [1]$$

Therefore, Abs is linearly related with k_υ (that is, at a given λ_0, with the atomic concentration of the analyte in the atomiser) and with the optical path of the light through the analyte atoms, L.

If the analytical or "line-method" of absorbance is used, as proposed by Walsh when using a hollow cathode lamp, the absorbance measurement is very simple and practical: the I_0 measurement needed (see equations above) is carried out virtually at the maximum, λ_0, of the absorption atomic line because the emission line from the hollow cathode lamp at the λ_0 will be much less broadened than the corresponding analyte absorption line in the flame (see Note insert below).

> While the hollow cathode lamp operates at low pressures and temperatures, and consequently its emission profiles, I_E, will not be broadened too much, a flame source burns at atmospheric pressure (760 torr) and at 2000-2500 K temperatures. In other words, the Lorentz and Doppler effects, respectively, will be much higher in the flame, and K_υ (the absorption line profile) more broadened, than in the hollow cathode lamp atomic clouds (see Figure 2.2 for visual illustration).

That is, the emission line halfwidth of the lamp is much smaller than the halfwidth of the corresponding absorption line in the flame, at the same wavelength ("lock-and-key" effect). This effect, a characteristic of the Walsh method, is illustrated in Figure 2.2, which shows how absorption coefficients measured will correspond to those at the maximum of the absorption line profile, K_0^ν . Therefore, K_0^ν value could be evaluated from the Doppler theory, which explains well the broadening in the centre of atomic lines observed in flames (as explained in Chapter 1).

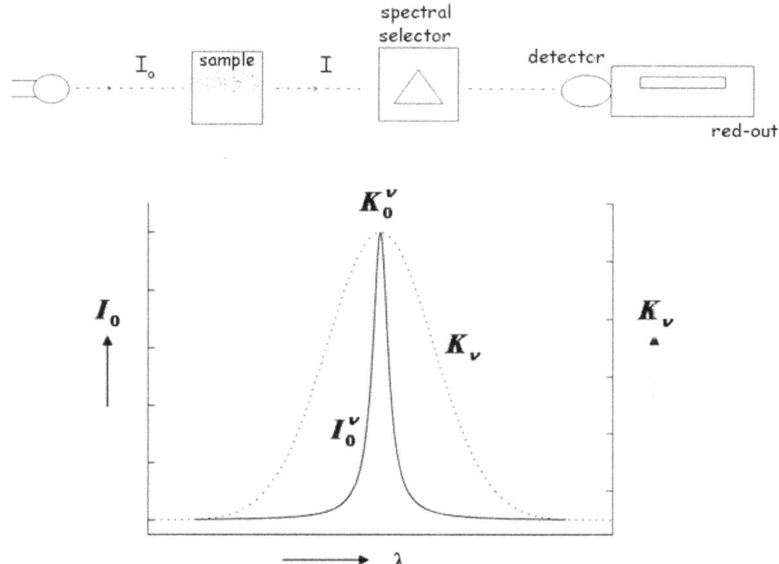

Figure 2.2. Basic AAS measurements and the "lock-and-key" effect using a hollow cathode lamp.

If the value of Abs is so measured at λ_0 maximum ("line-method"), the random thermal motion of analyte atoms relative to the observer is governed by a Maxwell distribution in the thermal equilibrium. Under such conditions, Doppler's theory demonstrates that K_υ distribution with wavelength can be expressed by a formula which at its maximum value, K_0^D, is given by the expression:

$$K_0^D = \frac{2\lambda_0^{\ 2}}{\Delta\lambda_D} \cdot \sqrt{\frac{\ln 2}{\pi}} \cdot \frac{\pi e^2}{mc^2} \cdot f \cdot N_0$$

where:

$\Delta\lambda_D$ = Doppler halfwidth of the absorption line used

e = charge of the electron

m = mass of the electron

c = light speed in vacuo

f = oscillator strength of the absorption line used

N_0 = analyte atom density (number of atoms per unit volume) in the ground state in the atomiser.

Now, if we introduce this value of K_0^D in the general expression of Abs we get:

$$Abs = 0,4343 \frac{2\lambda_0^{\;2}}{\Delta\lambda_D} \cdot \sqrt{\frac{\ln 2}{\pi}} \cdot \frac{\pi e^2}{mc^2} \cdot f \cdot L \cdot N_0$$

Thus, for a given set of experimental conditions in an absorption experiment, where most of the above magnitudes are constant, we get:

Abs = constant . L . N_0 **[2]**

and, at constant path-length or optical path L, absorbance is directly proportional to the analyte concentration in the atomiser.

2.4. QUANTITATIVE ANALYSIS BY ATOMIC ABSORPTION SPECTROMETRY.

Equation [2], derived above, for the "analytical" absorption of atoms in a flame, resembles very much the general Beer's law expression, considered as the theoretical basis of molecular absorption spectrophotometry, that is:

Abs = log $I_0/I = \varepsilon \,.\, l \,.\, C$ **[3]**

where "ε" is a dimensionless constant called *molar absorptivity* (or absorption coefficient), "l" is the optical path-length through the absorbing sample in the cuvette (in cm) and *C*, the analyte concentration ($mol.L^{-1}$).

Thus, the analogy of equations [2] and [3] is apparent, except for the way in which the analyte concentration is given: N_0 relates to that concentration in the atomiser (in the gas phase of the flame), while *C* refers to actual concentration of the absorbing compound (analyte) in the sample solution. Therefore, formula [2] may become formula [3], which is the mathematical expression of the fundamental law for absorption spectrophotometry, if the dependence between N_0 and *C* is clearly established.

This relationship between the atom concentration per unit volume ($N_T \approx N_0$) in the flame and the analytical concentration of the sought analyte, *C* (which is aspirated and nebulised from the sample solution into the flame, see Figure 2.3) can be approximated by an empirical formula of the type:

$$N_0 \cong N_T = \text{constant} \,.\, \frac{F.\varepsilon.\beta}{Q.T} .C \qquad [4]$$

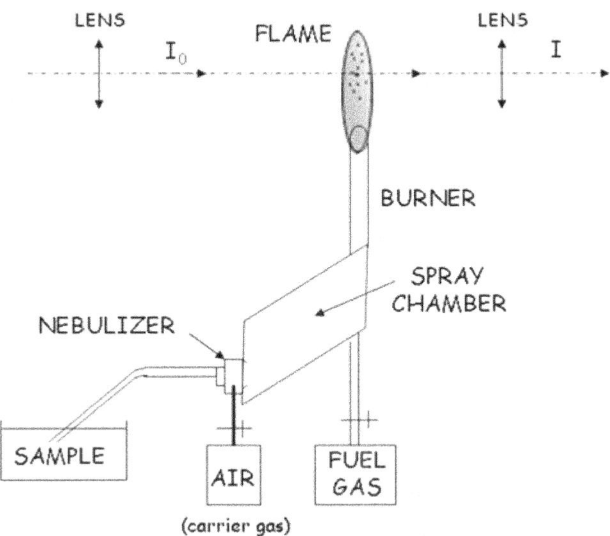

Figure 2.3. Classical sample introduction system for flame AAS measurements (via nebulisation of the dissolved sample).

where the different experimental variables (the values of which will be determined by the nebulization of the sample into the flame) determining the final atom population are as follows:

F = flow of the aspirated sample solution ($mL.min^{-1}$)

ε and β = efficiencies of vaporization and atomization processes in the flame, respectively.

Q = flow of the gases mixture in the flame (before burning, $mL.s^{-1}$)

T = absolute temperature in the atomiser (K).

Therefore, working in a laboratory under fixed nebulization conditions for a given atomic line of the analyte, it follows that we may say that N_0 is directly proportional to C in the solution ($N_0 \sim k'C$). In other words, if we combine [4] and [2] equations we obtain the formula:

Abs = k . k' . L . *C* = constant . L . *C* **[5]**

Thus, this basic equation of AAS is in fact a particular case of the general Beer's law of light absorption (see Z. Marczenco: "Spectrophotometric determination of elements". Ellis Horwood Ltd., Chichester, 1976, pages 6-10). In brief, from a practical point of view, by working in the laboratory at fixed experimental conditions including optical path, L (as it is customary in the practice of real AAS analysis) the measured analytical signal of Abs is directly proportional to the analytical concentration of the sought element in the sample solution. Of course, this "analytical signal" depends also upon other parameters (i.e. important experimental variables), which we have discussed to arrive finally at the formula [5] given above. For practical work in the laboratory all those experimental variables should be known, optimised for each analyte determination and eventually fixed accordingly. Then, the "calibration curve", that is the linear plot of "Abs" versus *C*, can be determined.

As this dependence may be complex, in practice we define the function of dependence experimentally for the fixed conditions of the analysis by measuring the Abs values observed for several standards of increasing and known concentrations of the analyte. Once the calibration curve (see, for example, Figure 2.4a) has been obtained experimentally, we may now obtain the values of C in any unknown sample solution, just by referring to such a calibration. This direct method is enough for many calibrations, but it is not valid in the presence of matrix effects (see the "multiplicative" interferences concept and how to tackle them in the box below).

> Once the blank values are subtracted from the total Abs measured, the linear relationships is given by: Abs = k.*C* and this is shown in the plot termed "aqueous standards" of Figure 2.4a.
>
> Sometimes, however, the matrix of a given sample may modify the slope (that is k, sensitivity) of this straight line obtained with pure aqueous standards. In such cases, direct calibration would produce critical analytical errors (due to the "so-called" multiplicative interferences) unless exact matrix matching is used for the calibration.
>
> Exact matrix matching is not always feasible as the precise matrix composition may be unknown or not completely known. In such cases the so-called "standard additions" method can be used: the sample is spiked with a few additions, five in the case shown in Figure 2.4b, of known amounts of the analyte in such a way that the sample matrix is not significantly changed. Then the absorbance of the unspiked and the spiked, appropriately diluted, samples is measured. As shown in Figure 2.4b, by extrapolation back to the negative side of the abscissa, the unknown and correct value of the analyte in the sample may be calculated (1.5 $mg.L^{-1}$ in that Figure rather than 1.1 $mg.L^{-1}$, as obtained using the Abs = k.*C* plot obtained for "aqueous calibration").

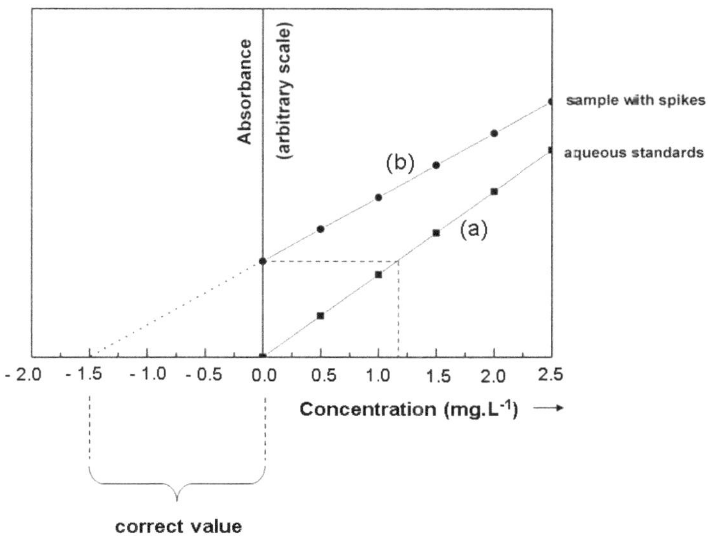

Figure 2.4. Main calibration modes in AAS. Conventional aqueous calibration. Standard additions calibration to correct for matrix interferences.

Of course, any slope change, observed between the slope for "aqueous standards" and the experiment with the analyte spikes added to the sample, is decisive proof of matrix or multiplicative interferences and so of the need of using standard additions for accurate determinations.

2.5. INTERFERENCES IN FLAME ANALYTICAL ATOMIC SPECTROMETRY TECHNIQUES.

In order to understand properly the exceptional analytical selectivity characterising flame-AAS methods, a basic knowledge of the different sources of interferences that may be encountered in atomic spectrometry is a must.

Revisiting the main interferences occurring using a flame as atomiser (in the three different measurement modes schematically shown in Figure 2.1) may reveal the general sources of error to be expected in the atomic spectrometry techniques. The main types of such sources of error are summarised in the Box below for the absorption, emission and fluorescence modes.

Main types of interferences (possible sources of error) in atomic spectrometry:

Affecting the three techniques (AAS, AES, AFS):

- Spectral
- Physical
- Chemical
- Ionization
- Variations of the temperature of the atomiser

Affecting absorption and fluorescence modes:

- Scattering of incident light

Affecting fluorescence (AFS) mode only:

- Fluorescence quenching

Thus, the concept and relative magnitude of each of the interferences will be described next and compared for the three main atomic modes, using the flame as the atomiser of reference. The following discussion is thus a sort of basic platform to understand and assess potential sources of error in any atomic technique we might use in our laboratory (i.e. concepts can easily be extrapolated to using an atomiser instead of a flame).

Spectral interferences

These are interferences brought about by the detection of a radiation that is due to or coming from elements (or molecules) different from the sought analyte. In other words, such "interfering" radiation spectrally overlaps the measured atomic line of the analyte within the used spectral bandwidth selected by the monochromator. Of course, a positive analytical error would be observed should such interferences occur.

Physical (transport) interferences

This source of interference is particularly important in flame-based methods, because it derives from the fact (see Figure 2.3) that the liquid sample must be aspirated and transported reproducibly, as a fine aerosol, into the flame.

Changes in the solvent, viscosity, density or surface tension in the sample solution will affect the final efficiency of nebulization and transport processes. Therefore, any of these changes or a deficient nebulization changing the effective transport of the analyte, as compared to nebulization and transport obtained initially for the standards used for calibration, will modify the final density of analyte atoms in the flame, even if its concentration in the sample is the same.

Chemical interferences

Using a comparatively low-temperature atomiser (as flames or graphite furnaces are compared to an inductively coupled plasma, ICP) these types of interferences are probably the most serious. Thus, their possible occurrence must be carefully assessed for accurate determinations.

Chemical interferences in atomic spectrometry are due to the presence or formation in the atomiser of analyte refractory compounds. Of course, when the temperature in the atomiser is not enough to dissociate and atomise completely the analyte a loss of the expected analyte atoms population will occur. The corresponding reduction of analyte atoms (those trapped in the refractory molecule) may bring about a considerable decrease of the atomic analytical signal.

Typical examples are phosphate interferences in blood serum determinations of Ca and Mg by flame-based atomic spectrometry methods. Phosphates of these metals can form and they are only partially dissociated and atomised at normal flame temperatures.

Another good illustration of the nature and importance of chemical interferences in flames is, for instance, the fact that elements such as Al, Si, Zr, Ta, Nb, etc. were thought, in the first years of AAS development, to be non-accessible to AAS determinations. For several years no atomic signals were obtained for such "refractory" elements in a common air/acetylene flame. In fact, only the introduction of "hotter" flames constituted by nitrous oxide/acetylene explained correctly the reasons for that, solving the required dissociation of the corresponding refractory oxides and hydroxides to form analyte atoms:

$$MO \xleftrightarrow{Energy} M^{o} + O$$

$$M(OH) \xleftrightarrow{Energy} M^{o} + OH$$

The above dissociations could not be achieved at around 2000 K with a conventional flame, but the higher temperatures occurring inside a N_2O/C_2H_2 flame (2800-3200 K) could dissociate the refractory oxides and hydroxides to provide free M^{o} atoms enabling strong, analytically useful, atomic signals for the determination of such elements.

Apart from using "hotter" flames and atomisers, an alternative and also efficient way to overcome chemical interferences is to resort to "releasing agents". These are chemical reagents (e.g. organic chelating compounds, such as 8-hydroxyquinoleine) that are able to form compounds with the analyte (competing with the refractory ones), which are easily dissociated in the usual temperatures of an analytical flame.

Ionization interferences

Particularly using "hot" flames, such as the N_2O/C_2H_2 flame, an opposite phenomenon which will, however, also reduce free atoms population may happen: if a low ionization energy metal, *M*, is introduced into a hot atomiser complete loss of the electron from the neutral atom is possible

$$M^o \xleftrightarrow{+Energy} M^+ + e^-$$

in such a way that ionization may take place and the M^o population will decrease (that is, the sensitivity of the analyte determination, for which an atomic line of M^o is used, is proportionally reduced).

Fortunately, these ionization interferences can be suppressed by adding to the sample solution another element or compound which can provide a great excess of electrons in the flame (i.e. another easily ionisable element). In this way the above ionization equilibrium is forced to the left in the presence of the excess of electrons introduced in the flame.

Well known examples of the use of such buffering compounds are salts of Cs and of La (easily ionizable elements) which are widely used as ionization buffers in the determination of metals such as Na, K or Ca by flame-AAS (or flame-OES).

Temperature variations in the atomiser

Variations in the temperature of the atomiser would change the population of neutral atoms (M^o) needed for atomic absorption, but particularly of excited atoms (M^*), essential for atomic emission measurements. Therefore, the use of atomisers for which a constant temperature is maintained during operation should be aimed at in order to secure reproducible analytical signals.

Light scattering and unspecific absorptions

Both types of problems are only seen when an external line source is used (that is, as Figure 2.1 shows, just in AAS and AFS measurements). When part of the light coming from the hollow cathode lamp, I_0, is scattered by small particles in the flame (e.g. droplets or refractory solid particles) or perhaps absorbed unspecifically (e.g. by undissociated molecules existing in the flame) important analytical errors would be derived if no adequate corrections are made of such unwanted phenomena: firstly, because the scattered or dispersed radiation diminishes the real hollow cathode lamp intensity, I_0, affecting the measurement. What is perhaps more important, false analytical signals could be measured and so confused with the specific absorption due to the analyte atoms. That is, if dispersed radiation or unspecific molecular absorptions are detected, these signals are not due to the analyte and would result in gross errors.

Fortunately, both sources of "false signals" can be easily distinguished from the "specific" analyte signals, which do occur at the analytical line position only (and not outside it in the spectrum). This basic differential feature can be adequately used for correction (see "background correction" in Chapters 3 and 6).

Quenching of the fluorescence

For the sake of completeness it is adequate to finish this section by referring to a type of interference that is only of interest in the fluorescence type of measurement (Figure 2.1b). Quenching is a deactivation of excited states in atoms (or molecules) due to collisions with the surrounding atoms or species existing in the atomiser (e.g. coming from the burning flame or the matrix of analysed samples). This deactivation is non-radiational and so it will decrease the intensity of the fluorescence observed.

This fluorescence quenching can be minimised by using an inert gas environment for the analyte atoms. If an inert noble gas surrounds the excited atoms quenching is very small as compared with molecular gases. Thus, flames "purged with Ar" have been recommended for flame atomic fluorescence.

2.6. COMPARATIVE ANALYTICAL PERFORMANCE CHARACTERISTICS OF AAS

As pointed out already, the most popularly, used and sold experimental arrangement of the three possible measurement modes illustrated in Figure 2.1 is, still today, the one featuring those components and the arrangement characterising an atomic absorption spectrometer. From the experience and practical perspective gained by more than half a century of routine work with these flame-based atomic techniques, this popularity must be linked to the problem-solving capability that each of the three arrangements offers to the analytical chemist. In other words, in the long run the eventual success of one or more of these three techniques (AAS, AFS and AES) is determined by the intrinsic analytical performance characteristics of each of the measurement modes.

It is true that the three approaches are in a way complementary for application to solving problems requiring elemental analysis information. As explained in Section 1.5 of the first chapter "there are horses for courses" (i.e. we will need different techniques for different analytical problems) and so we should not purchase an atomic technique for dissolved sample analysis if the main requirement of our laboratory is direct solids analyses.

Thus, it may not always be clear which atomic technique is optimum to tackle all (or most of) the analytical problems faced in a routine laboratory. However, it is perfectly clear that to decide on that best suited for your particular laboratory you need two key pieces of knowledge: first, the real problems and elemental analysis type to be faced must be well defined, and second, once the analytical requirements in your laboratory have been clearly established, knowledge of the analytical performance and capabilities of the different available atomic techniques is mandatory for an effective selection.

In order to be able to compare the differing analytical techniques, the use of some critical parameters, worldwide accepted as key criteria for comparisons, is now commonplace.

Sensitivity, detection limits, analytical working range, selectivity, accuracy and precision, cost, sample throughput and availability of well-proven methodologies are key criteria and their relative importance must be clear to the eventual user of those techniques. Using the same atomiser (e.g. a flame which is still the most popular one) the techniques derived from the three basic designs in Figure 2.1 exhibit fundamental advantages and limitations, which should be properly understood.

Therefore, these comparative key criteria and their characteristics for the three flame techniques will be now comparatively discussed, since such discussion is of fundamental importance, not only to understand the flame techniques but also to introduce fundamental concepts (which can be extrapolated to any instrumental analytical technique) for establishing eventual analytical performance characteristics and methods' validation and comparisons.

Sensitivity and detection limits

Analytical sensitivity, S, refers to the slope of a linear calibration graph in an instrumental method of analysis. In AAS methods, sometimes the term "sensitivity" means "the concentration of the analyte producing an Abs=0.0044 (1% absorption)", which is a form of assessing the value of S.

However, to clearly define the detection power of a given technique or method, that is the lowest detectable concentration of the analyte, a more general concept is used: the detection limit (DL) of the technique or method for that particular element. The detection limit informs us of the minimum concentration (or amount) of the analyte that can be detected, with a given level of certainty, by the considered method. Based on statistical considerations (taking into account the degree of certainty required) it has become customary to define the DL following the IUPAC criterium: the DL of any analytical method refers to "the concentration (amount) of the analyte producing a net signal three times greater than the value of the observed standard deviation of the blank (i.e. the sample without analyte) signal", σ_B. In other words, C_{DL} means the minimum concentration value of the analyte in the sample, which produces a signal distinctly higher, with a known probability, than that observed for the blank.

Thus, at the C_{DL} level and in Abs measurements:

$$\boldsymbol{Abs_T^{DL} = Abs_{blank}^{DL} + 3\sigma_B} \quad \text{or} \quad \boldsymbol{Abs_{net}^{DL} = 3\sigma_B}$$

(being $\boldsymbol{Abs_{net} = Abs_T - Abs_{blank}}$)

As the sensitivity is given by the slope of the calibration graph, $\boldsymbol{S = \frac{Abs_{net}}{C}}$,

at any given analyte concentration, we have:

$$\boldsymbol{S = \frac{Abs_{net}}{C} = \frac{Abs_{net}^{DL}}{C_{DL}}}$$

And so a good estimation of the "detection limit" (C_{DL}) is:

$$C_{DL} = \frac{Abs_{net}^{DL}}{Abs_{net}/C} = \frac{3\sigma_B}{S} \qquad \textbf{[6]}$$

This formula shows that C_{DL} is better (lower) not only when the S of the technique is higher but also when the standard deviation associated with the blank, σ_B, is smaller.

Sometimes the term "determination limit" or limit of quantification (L_Q) is also used to express the minimum concentration needed for a reliable determination of the analyte. Again, based on statistical considerations on the acceptable uncertainty of the value given, the valuable parameter is defined by IUPAC as $L_Q = 10\frac{\sigma_B}{S}$.

Probably the C_{DL}s are the most common approach to compare several analytical techniques (or methods) from the point of view of their detection power for a given analyte. Unfortunately, C_{DL} values in the literature are often misleading (usually overly optimistic). Perhaps for practical purposes in the laboratory L_Q values may provide a more realistic figure in method evaluations.

It is important to have a relative picture of the whole typical detection limit ranges for the most important atomic spectrometry techniques available today for routine elemental analysis. Figure 2.5 provides such a picture for the three AAS techniques considered in this book, as compared to more sophisticated ICP-based atomic techniques. It is clear that the best detection limits are provided by ICP-MS, but graphite furnace-AAS and hydride generation-AAS may offer in some cases extremely good detection limits at a comparatively much lower cost.

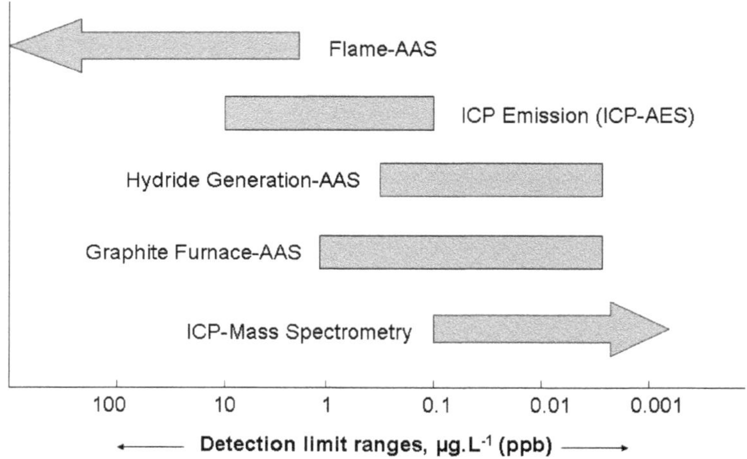

Figure 2.5. Typical detection limit ranges for the most important atomic spectrometry techniques.

For flame-based atomic techniques, flame-AAS provides the most versatile technique, as it is applicable to the widest range of elements over a wide range of concentrations. Flame-AES can be appropriate for alkali and alkaline-earth metals, while flame-AFS may provide the best sensitivity for others (including metals of particular clinical interest such as Hg, Cd, Pb or Tl).

Selectivity of the three flame-based techniques

As pointed out above, it is useful to discuss the comparative selectivity of each technique by careful examination of the extent of the different types of interferences, described before, expected in each technique. Thus, a brief revision of the relative magnitude of these interferences in each case is given below.

When considering spectral interferences it is fair to start by stressing that atomic spectrometry provides much better spectral selectivity than corresponding UV-VIS molecular absorption spectrophotometry methods (atomic line halfwidths are orders of magnitude narrower and so spectral overlapping will be much less likely).

When the magnitude of these interferences are compared for emission, absorption and fluorescence atomic spectrometry it can be said that in general terms the so-called "lock-and-key" AAS effect (Fig. 2.2 shows that the emission of the HCL is the "key" entering precisely the "lock", that is the absorption profile of the analyte) provides the most selective mode; working with a hollow cathode lamp real line spectral interferences are virtually non-existent, even using a relatively low resolution monochromator. The atomic emission (Figure 2.1c) arrangement should be more prone to spectral overlapping (in practice this is particularly true if the temperature of the excitation source is high, e.g. in ICP-OES, and many energy levels can be excited). Finally, regarding flame-AFS it can be said that using a hollow cathode lamp for excitation specific spectral interferences are rare in atomic fluorescence, because only the emission lines of the analyte should be brought about by its atoms excitation in the flame. In any case, when "resonance" fluorescence is used for the analysis (i.e. the same frequency is used for excitation and for the final fluorescence measurement) light scattering of I_0 in the flame would produce a most serious spectral interference at the AFS measurement frequency. Of course, such emission is unspecific and could be corrected for by using background correctors (based on measuring the unspecific background, away from the fluorescence atomic line, as we will discuss later on for AAS measurements).

Physical interferences, affecting the efficiency of the analyte transport to the flame, will usually decrease the observed AAS signal. Although less analyte is usually transported (e.g. because samples have higher viscosity than the standards used for calibration) the reverse effect is also possible: if the formation of volatile compounds takes place during nebulization (e.g. for osmium, the formation of volatile O_sO_4 in the presence of HNO_3 acid) AAS enhanced signals may be also observed. In any case, absorption, emission or fluorescence modes will be equally affected by such alterations.

Chemical interferences are known to be most serious in a flame due to its comparatively low temperature. In either case, as the three modes depicted in Figure 2.1 need an efficient atom formation, the expected analytical selectivity deterioration, due to refractory compounds formation in the flame, will be identical for the three measurement modes.

The case of **ionization interferences** is analogous: if the measured atoms, M^{o}, disappear to form ions, M^{+}, the analytical signal will decrease but the observed magnitude of the decrease will be again identical for the three flame-based techniques shown schematically in Figure 2.1 (i.e. all of them rely on efficient and reproducible analyte atoms formation).

Variations in the atomiser temperature, T, will, of course, affect the AAS measurements. However, such variations will have an immediate effect on the population of the atoms in the excited state 2, N_2 (i.e. we know that $\frac{N_2}{N_0} \cong e^{-\frac{h\nu}{kT}}$, that is, the magnitude of N_2 depends exponentially on the absolute temperature in the flame). Therefore maintaining a constant T during measurements is critical in flame-atomic emission methods.

For AAS measurements the important parameter is N_0 (the atom population at the fundamental level) and N_0 is barely influenced by small variations (e.g. ≤ 2 %) in the flame temperature. In brief, errors derived from fluctuation in the flame temperature are less important in AAS than in AES measurements.

Light scattering interferences do not exist in emission, but they may be a source of analytical errors in both AAS and AFS because we have to use an external light source of excitation (the lamp). Fortunately, all Rayleigh light scattering interferences are unspecific in nature (i.e. while they occur at the atomic analytical line, they also occur both close to and away from the measurement wavelength). Thus, such problems are routinely tackled in an automatic manner using an adequate "background corrector" system, incorporated today as an accessory of modern AAS spectrometers. This, as said before, is a similar problem in AFS and so appropriate steps should be taken to correct such possible sources of analytical errors.

Accuracy and precision

It is known that the accuracy of analytical methods relates to the observed closeness between the "measured" value (the mean of a series of replicates, obtained by repeating the same analytical determination a known number of times) and the "true" value of the sought concentration (or amount) of the analyte. The difference between the two values gives us the analytical error. For AAS the main sources of such error were detailed in Section 2.5. That is, the presence of interferences will bring about a "bias" of the measured value in a given direction (i.e. a positive interference will originate always positive errors).

However, even if a most careful analyst repeats the same determination many times some variability of the obtained results is observed. This variability depends on many uncontrolled factors (it is hazard-dependent and so unavoidable to a certain extent) but it can be addressed by the repeatability "standard deviation", SD, observed in the "n" repeats carried out, $SD = \sqrt{\frac{\sum (x_i - \bar{x})^2}{n-1}}$

where x_i is the value from any individual measurement and $\bar{x}$ the mean value for all the repeats carried out.

The relative standard deviation is then, $SD_r = \frac{\sigma}{\bar{x}}$; it is usually expressed as a % and termed "coefficient of variation", $SD_r(\%) = \frac{\sigma}{\bar{x}}.100$, of the mean.

Of course, appropriate and careful control of interferences is mandatory in order to avoid "bias" in AAS determinations. Moreover, the repeatability of our AAS measurements should be also optimised and eventually worked out.

Typical concentration levels determined by flame-AAS methods are in the range of a few mg.L^{-1} and repeatability SD_r (%) values observed for such concentrations are always better than ± 1%. This value is clearly lower (better) than the corresponding SD_r (%) observed using AES or AFS measurements.

Analytical linear range

The linear relationship between measured Abs and the sought analyte concentration, *C*, is governed by Beer's law (equation [5]), but the range of *C* values between which such a relationship holds is not very large. There are deviations from the linear behaviour due to several factors, particularly when *C* values are high (typically three orders of magnitude above the corresponding limits of quantification). This particular parameter is much less troublesome with AES and AFS methods, where five or six orders of magnitude of analytical linear ranges are commonly achieved.

In brief, the analytical working range can be viewed as the concentration range of the analyte in the sample over which quantitative results can be obtained without further recalibrations. As mentioned, AAS offers, generally speaking, more limited ranges than emission-based methods.

Versatility and sample throughput

About 70 elements of the Periodic Table can be determined at 0.1-100 ppm levels using modern flame AAS methods. However, AAS techniques are basically single-element techniques (this is particularly so for graphite furnace-AAS, where the lamp and the atomisation conditions have, as a rule, to be selected individually for each element). This feature determines that sample throughput in AAS (specially with electrothermal atomisation, as detailed in Chapter 6) is comparatively low.

Atomic emission techniques (particularly those using a hot spectrochemical source as in ICP-OES) are intrinsically multielemental and this characteristic offers the possibility of a very high sample throughput for routine analyses. To counterbalance such a clear advantage, however, AAS techniques are simpler and so considerably less costly than ICP-OES or ICP-MS techniques (which do offer high throughput multielemental analysis).

Robustness and availability of well-proven methodologies

To conclude this chapter it is fair to say that flame-AAS is probably the easier-to-use and more robust of the three atomic techniques shown schematically in Figure 2.1. Extensive information on analytical methodologies and protocols for elemental determinations by flame-AAS are available from the AAS instruments manufacturers (see Chapter 8 for details on ten such companies).

The AAS arrangement of Figure 2.1a provides a comparatively more robust measurement, because the analytical signal measured is Abs = $-\log I/I_0$. That is, Abs is a ratio of two measurements of intensity, rather than an absolute emission intensity measurement, typically required for the emission and fluorescence techniques. This ratio method will cancel out many sources of variability affecting a single measurement.

In brief, AAS techniques may provide most robust analytical procedures to the point of representing today the more reliable atomic technique for elemental determinations at the appropriate concentration level (see Figure 2.5 for an overall comparison of concentration ranges).

Chapter Three

BASIC COMPONENTS OF ATOMIC ABSORPTION SPECTROMETRIC INSTRUMENTS

3.1. INTRODUCTION: SINGLE BEAM AND DOUBLE BEAM INSTRUMENTS

An atomic absorption spectrometer consists basically of three main parts: a light source, a sample cell in which gaseous atoms are produced, and a means of measuring the specific light absorbed. Figure 3.1a shows the simplest configuration, called a "single beam spectrometer"; as can be seen it comprises, a lamp that emits intense and narrow lines, the atomizer where the species of interest are converted into atoms, and the wavelength selector (e.g., a monochromator) to select the specific wavelength of light - i.e. spectral line – which is absorbed by the analyte, all of which are aligned. The light wavelength selected is directed onto a detector, e.g. a photomultiplier tube, producing a signal proportional to the light intensity. The combination of the wavelength selector with a photoelectric detector of the isolated wavelength band(s) is usually called a spectrometer. However, it is important to note that in absorption spectrochemical measurements the term "spectrometer" refers to the whole instrumentation, containing also the lamp and the sample cell.

a)

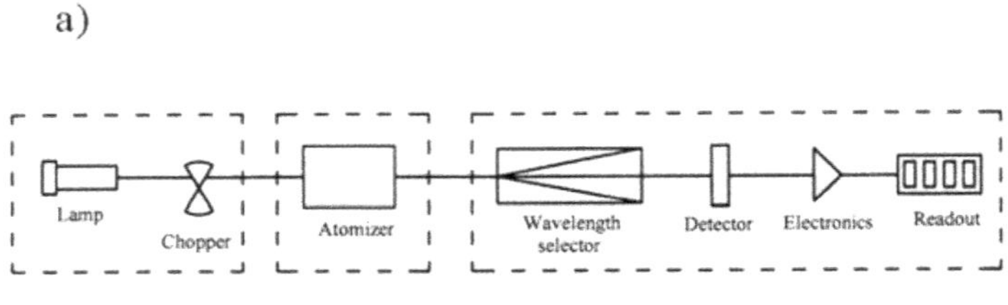

b)

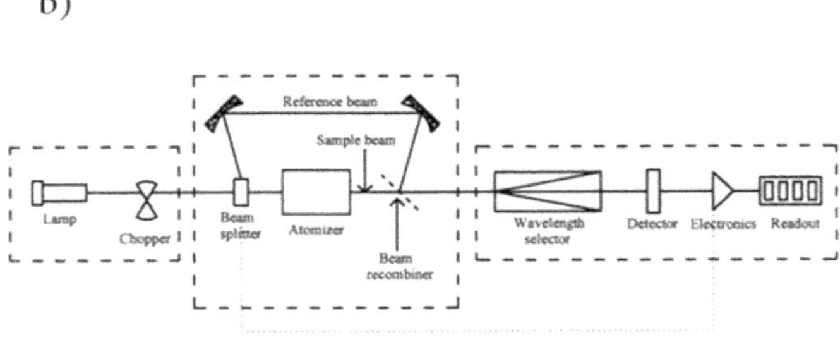

Figure 3.1. Schematics of instruments for atomic absorption spectrometry. a) Single beam spectrometer. b) Double beam spectrometer.

To eliminate the atomiser continuum emission, the radiation source is modulated to provide a means of selectively amplifying modulated light coming from the source lamp, while continuous emission from the sample cell is disregarded. Source modulation can be accomplished with a rotating chopper (mechanical modulation) located between the light source and the atomizer, or by pulsing the source with a pulsed power supply (electronic modulation). A synchronous detection eliminates the unmodulated DC signal emitted by the atomizer and so measures only the amplified AC (modulated) signal coming from the lamp.

Unfortunately, the intensity of the light source may not remain constant during the time of analysis. If only a single beam (Figure 3.1a) is used, a blank reading containing no analyte would need to be taken first, in order to set the absorbance readout to zero. Of course, if the intensity of the source changes by the time the sample is put in place, the recorded absorbance will be inaccurate. To obviate such an inconvenience, an alternative instrument configuration is based on a spectrometer that incorporates a beam splitter so that one part of the beam passes through the sample cell and the other is used as a reference. A diagram of a "double beam spectrometer" is shown in

Figure 3.1b. In a double beam instrument there is a continuous monitoring between the reference beam and the light source signals. To ensure that the spectrum does not suffer from loss of sensitivity, the beam splitter is designed so that a proportion as high as possible of the total energy of the lamp beam is passed through the sample.

In the following sections of this chapter the choices for the basic components of an atomic absorption spectrometer will be considered: lamps, atomizers, wavelength selectors and detectors. It should be noted that any radiation from the light source absorbed or scattered by other atoms or molecules other than the analyte will give rise to a background absorption which will add to the specific absorption of the analyte. Thus, in the last section of this chapter the means for measuring and correcting for this background absorption will be dealt with.

3.2. PRIMARY RADIATION SOURCES

By far, the most common lamps used in AAS are sources emitting a narrow-line spectrum of the element of interest. This line should be of sufficient spectral purity and intensity to achieve a linear calibration graph and low level of baseline noise. If the intensity of the spectral line is too low, the noise performance of the instrument will be compromised by excessive photon shot noise (this arises from the random generation of electrons in the UV-Vis light detector). If there is unspecific non-absorbable radiation within the spectral band pass of the wavelength selector, the calibration curve will be non-linear and curve towards the concentration axis, bringing about a loss of sensitivity at high absorbance values. Such non-absorbable radiation can result, for example, from the presence of another spectral line within the spectral band-pass of the monochromator or from a continuum background radiation.

The main sources used for AAS are the hollow cathode lamp (HCL) and the electrodeless discharge lamp (EDL). The HCL is a bright and stable line emission source commercially available for most elements. However, for some volatile elements (where low emission intensity and short lamp lifetimes are commonplace), EDLs are used. This is the case for As, Se, Hg, etc, AAS lamps. In addition, boosted HCLs aimed at increasing the output from the HCL, are also commercially available.

Emerging alternative sources, such as diode lasers and the combined use of a primary source emitting a continuum and a high resolution spectrometer will also be reviewed in the following discussion.

3.2.1. Hollow cathode lamps

A diagram of a typical hollow cathode lamp is shown in Figure 3.2a. It consists of a hollow cathode (containing the element of interest) and an anode. These are sealed in a glass tube filled with and inert gas at low pressure. By applying a potential difference of about 300-400 V between the anode and the cathode, the ionisation of gas occurs and so a discharge between two electrodes takes place. At the low pressure, positively charged gas ions (e.g. Ar^+, Ne^+) are accelerated towards the cathode by the potential existing in the discharge. Under impact with the solid, metal atoms from the cathode are ejected, in a process called "sputtering" (see Figure 3.2b), to the plasma existing at the mouth of the cathode. Once in the plasma, sputtered metal atoms may collide with other high-energy particles, resulting in a transfer of energy and causing the metal atoms to become excited. Since this excited state is not stable, the metal atoms relax to their ground state, emitting radiation at the characteristic wavelengths of the element. In this way, more than one analytically useful spectral line can be generated for most elements.

The cylindrical shape of the cathode concentrates the radiation into a beam, which passes through a transparent window.

a)

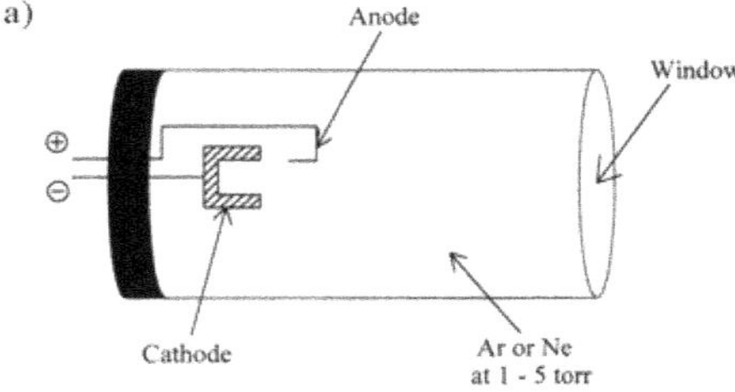

b)

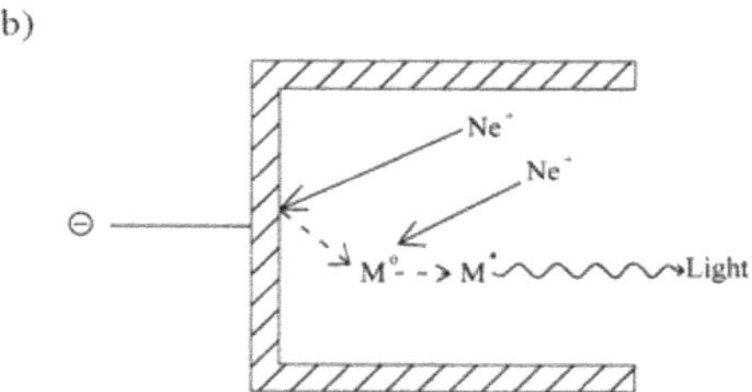

Figure 3.2. The hollow cathode lamp.

a) Diagram of a hollow cathode lamp.

b) Sputtering, excitation and emission processes.

3.2.1.1. *The components of the hollow cathode lamp*

Anode and cathode: the HCL contains a tungsten or a zirconium anode and a cylindrical hollow cathode made from, or containing, the element to be determined. If the metal is stable in air and has a high melting point, the pure metal may be used (e.g. cathodes for lamps of Cu, Fe, Ni and Al are usually machined from the corresponding solid metal, while hollow cathodes for expensive metals such as Pd, Au and Ir usually have sheet metal inserts). If the metal is too brittle, a sinter of pressed metal powder is used (e.g. Mn, W). If the metal is reactive in air, then the metal oxide or halide can be used (e.g. Na). A metal with a relatively low boiling point or a relatively high vapour pressure (e.g. Hg, Cd, Pb, Zn) is usually alloyed with another metal. The powder technique is also used for producing multielement lamps in which two or more metals are present (see section 3.2.1.3).

Gas fill: the filler gas must be monoatomic in order to avoid molecular continuum spectra. Either high-purity neon or argon (1 - 5 torr) is used as the discharge gas. Generally, argon is only used when a neon spectral line would interfere with the wavelength for the selected element. The preferential use of neon is due to its higher ionisation potential which produces greater signal intensity. Helium is not used due to its low mass number making it an inefficient sputterer. Besides, helium gives a short lamp lifetime due to a rapid lowering of the filler gas pressure by adsorption of gas atoms onto surfaces within the lamp.

Envelope: the electrodes are enclosed within a glass envelope with a quartz, or a special borosilicate, end window attached to it. For elements that emit at wavelengths of less than about 300 nm, quartz must be used. For higher wavelengths borosilicate is typically used.

3.2.1.2. *Hollow cathode lamp operation*

HCL electrical current is the key parameter affecting the analytical AAS results. The principal effect of an increase in the lamp current is an increase in the intensity of the lamp element-specific light emission. However, as the operating current increases above the recommended value, increased broadening of the emission lines occurs and the phenomenon of self-absorption is observed at very high HCL currents. In the presence of temperature gradients an inversion of the peak top of the emission line (due to a cooler cloud of atoms in front of the cathode absorbing the cathodic emission within the lamp itself) may occur. Such "self-reversal" effects gives rise to reduced sensitivity and, for some elements, a shorter linear interval on the calibration curve. However, it is worth mentioning that for measurements near the detection limit using lamp currents higher than that recommended may be advantageous.

From a practical point of view one must ensure a good stability of the hollow cathode lamp signal. Typical HCLs require a warm-up period, (some minutes) after switch on, to stabilize their light output. Warm-up time is particularly important for single beam instruments (the change in intensity of the lamp is reflected in the baseline of the instrument). With double beam instruments, the need for lamp warm-up time is not so critical, since the instrument is able to compensate for changes in the sample beam intensity by making continual comparison with the reference beam. Nevertheless, it is desirable to allow a short warm-up before attempting precise analytical measurements. This is because the profile of the emission line from the lamp can change during this period and so small changes in analytical signal may result.

Even the most carefully handled lamp will eventually fail to operate, as the filling gas is absorbed into the internal surfaces of the glass lamp. In time, the pressure of the fill gas will fall to a level that can no longer sustain the hollow cathode discharge. Operating the HCL at excessive lamp currents will accelerate this process. In addition, attempts to run a lamp at extreme currents can cause the cathode to overheat, and this can damage irreversibly the cathode (this is especially true for the more volatile elements).

For some marketed HCLs it can be observed that brand-new lamps look like they have been already used (see Figure 3.3). Two treatments during processing in the lamp production step account for this:

- some manufacturers heat treat the cathode material under vacuum to ensure that all the absorbed gases are removed. During this purification stage (aimed to ensure a dependable performance), a layer of the cathode material is deposited on the inside of the glass envelope of the lamp. The amount of material deposited varies, depending on the volatility of the cathode element.

- also, a characteristic black patch on the lamp envelope near the anode can sometimes be seen, which is produced by subjecting the zirconium anode to ion bombardment. This step vaporizes a small amount of anode material and deposits it on the lamp envelope producing the black spot. This zirconium metal film is highly reactive and acts as a very efficient getter of traces of oxygen and other impurity gases that might otherwise reduce the lifetime of the lamp. Therefore, the black getter spot close to the anode helps to prolong the useful life of the lamp and ensures continued spectral purity throughout the life of the lamp.

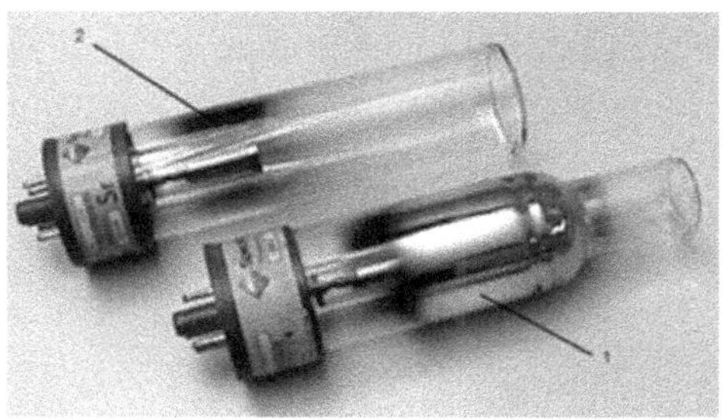

Figure 3.3. Appearance of some brand-new lamps. The thin layer of cathode material deposited onto the inside surface of the glass envelope (1) is noticeably different to the black getter spot (2). Reproduced with permission of Varian.

3.2.1.3. *Multi-element hollow cathode lamps.*

The HCL described above is designed to emit the atomic spectrum of a desired single element. Thus, a specific lamp should be selected for each element to be determined. However, multi-element HCLs are also commercially available. In practice, multi-element HCLs are not widely used because not many of them fulfil the following requirements:

i) Spectral interferences should be absent. Spectral interference is the overlap of an emission line of the analyte with the emission line of the other(s) element(s) (interferent(s)). For example, proposed Ag/Au lamps failed in this respect, since emission lines from the silver interfere with the analytical line of gold and, in turn, gold emission lines interfere with the silver analytical line.

ii) The chosen elements should be compatible, i.e. they should co-exist in the same cathode without undesirable effects restricting the lifetime of the lamp.

iii) The selected analytical line for each element has to provide enough intensity, while keeping a low baseline noise.

iv) The multi-element HCL must satisfy a market need. For example, a Ag/Cd/Pb/Zn lamp is useful for the environmental analysis market, while the Co/Cr/Cu/Fe/Mn/Ni lamp meets the needs of the base-metal market.

3.2.2. Electrodeless discharge lamps

For most elements the hollow cathode lamp is the most satisfactory spectrochemical source for atomic absorption. However, for volatile elements, such as As and Se, the low intensity and short lamp life are a problem when using HCL designs. For such elements, electrodeless discharge lamps offer recognised advantages in terms of useful lifetime of the lamp and higher light intensity.

The basic design of an EDL is shown in Figure 3.4. A small amount of the metal or salt of the element for which the source is to be used is sealed inside a quartz bulb, which contains argon at low pressure. A radiofrequency (or microwave) coil surrounds the bulb. When power is applied, an intense radiofrequency (or microwave) field is created. The argon within the tube ionises and gains kinetic energy from the field forming a low pressure plasma. The plasma energy is transferred to the vaporised metal upon collisions. Finally, the excited metal vapour returns to its ground state by emitting light. An auxiliary power supply is necessary to operate EDLs with most spectrometer models.

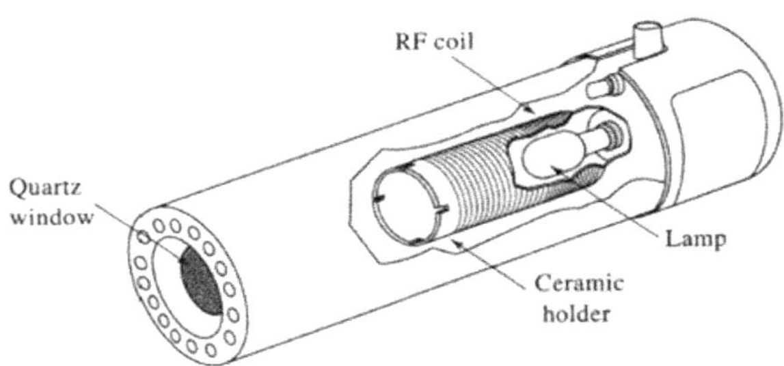

Figure 3.4. Schematic of an electrodeless discharge lamp. Reproduced with permission of Perkin Elmer.

The EDLs are about ten times more intense than HCLs, but they usually suffer from unstable output. EDLs offer the analytical advantages of better precision and lower detection limits if the determination is intensity-limited. Finally, electrodeless discharge lamps are available for a wide variety of elements, including arsenic, antimony, selenium, mercury, cadmium, bismuth, phosphorus.

3.2.3. Boosted discharge lamps.

The boosted discharge lamps are HCLs in which a second discharge, electrically isolated from the sputtering discharge, is used to further excite the sputtered atoms. In the conventional HCL the only way to increase the intensity of the emitted light is to increase the lamp current. This causes more atoms to be sputtered leading to curvature of atomic absorption calibration graphs, because the radiation from excited atoms in the cathode can interact with unexcited atoms on their way out of the cathode, giving rise to self-absorption and self-reversal of resonance lines.

In the boosted discharge lamps the excitation process is made independent of the sputtering process, thus allowing a more efficient excitation of the sputtered atoms. In this way, the radiation does not have to pass through a cloud of ground-state atoms on its way out of the lamp; consequently, there is less self-absorption and the calibration graphs are less curved. Also, a preferential excitation of the lines of analytical interest is observed in many cases.

However, boosted discharge lamps are not equally effective for the interesting analytical lines of all elements. In fact, in some cases little or no increase (and even for some particular elements a reduction) in intensity of the corresponding analytical line have been observed when boosting a HCL.

3.2.4. Diode lasers

The potential of diode lasers as sources of resonance radiation for AAS has been increasingly investigated in recent years. Single mode diode lasers are used in AAS as tunable-narrow band line sources, offering a number of attractive features:

i) The power of presently available commercial diode lasers is between one and several orders of magnitude higher than that provided by HCLs. In addition, diode lasers show good stability, both in terms of wavelength and intensity.

ii) In contrast to the line sources described above, a diode laser emits a prominent single line, under normal operating conditions, which dramatically simplifies the spectral isolation of the absorption signal.

iii) The typical line width of a commercial diode laser is approximately two orders of magnitude less than the width of absorption lines in flames and furnaces. This allows the expansion of the linear dynamic range of the calibration curve to high concentrations of an analyte, by detection of absorption in the wings of the absorption line.

iv) The spatial coherence of diode lasers makes it possible to deliver a narrow laser beam at a distance without noticeable divergence and to easily manipulate the spatial profile.

v) The wavelength of diode lasers can be easily modulated at frequencies up to GHz by modulation of the diode current. Wavelength modulation of the diode laser with detection of absorption at the second harmonic of the modulation frequency greatly reduces low frequency noise (flicker noise) in the baseline and provides improved detection limits.

Unfortunately, perhaps the main drawback of present diode lasers for AAS is the narrow spectral range of the commercially available diode lasers, which limits their applicability drastically when compared with HCLs.

3.2.5. Continuous sources

Line sources, such as HCLs and to a lesser extent EDLs, have been used almost exclusively for routine application of AAS. Their stable, narrow-line emission at the center of the absorption profiles (as proposed by A. Walsh originally) guarantees high analyte specificity and good detection limits, even with the use of low-resolution monochromators. Conversely, the use of a source emitting a continuum will require a spectrometer providing a resolution as high as approximately 2 pm (picometers). The present availability of continuum sources with a high emission intensity within the spectral interval of interest, together with high-resolution echelle spectrometers and solid-state arrays detectors, have prompted the exploitation of the advantageous use of continuous spectrum light sources in AAS.

A high intensity continuous source (a xenon short-arc lamp) in combination with a high-resolution double-echelle monochromator and a linear CCD array detector has been commercialised for AAS measurements. Figure 3.5 shows the experimental set-up for such an AAS arrangement using an electrothermal atomizer. Four major advantages of the configuration are:

i) improved signal-to-noise ratio because of the high intensity of the radiation source. In addition, the atomic absorption can not only be measured at the centre of the absorption line (with maximum sensitivity) but also at its wings (with reduced sensitivity), thus greatly increasing the dynamic range.

ii) the entire spectral environment around the analytical line becomes "visible", giving more information than common AAS instruments, which eventually results in a more reliable and accurate background correction.

iii) new elements might be determined, for which no radiation source is available.

iv) the system makes possible a truly simultaneous multi-element AAS measurement by replacement of the one-dimensional array detector by a two-dimensional multi-array detector (as it is common practice in optical emission spectrometry).

The practical use of these AAS spectrometers, however, is still comparatively low.

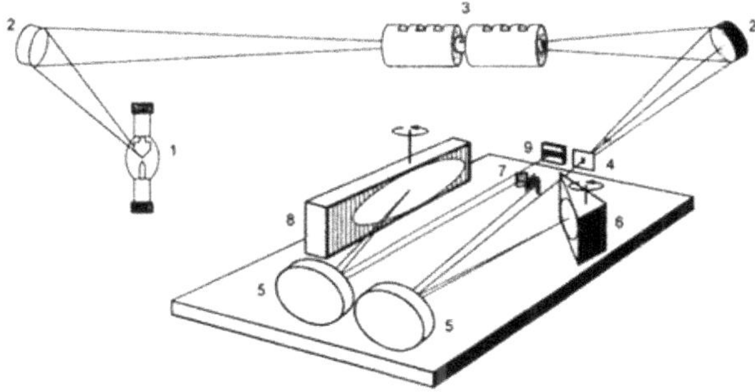

Figure 3.5. Experimental set-up for continuous source AAS with double echelle monochromator. (1) Xenon short-arc lamp, (2) off-axis ellipsoid mirrors; (3) longitudinal Zeeman graphite furnace module, (4) entrance slit, (5) off-axis parabolic mirrors, (6) Littrow prism, (7) deflection mirrors and intermediate slit, (8) echelle grating (76 grooves/mm; blaze 76°), and (9) linear CCD array detector (from U. Heitmann, M. Schuetz, H. Becker-Ross, S. Florek, Spectrochim. Acta Part B 51 (1996) 1095-1105). Reprinted with permission of Elsevier.

3.3. ATOMIZERS: OVERVIEW

The most common form of samples in AAS is as "dissolved" samples. Two systems are most frequently used in AAS to produce atoms from a *liquid* or *dissolved* sample:

i) a flame, where the solution of the sample is aspirated. This system is addressed in Chapter 4 of this book dedicated to flame atomic absorption spectrometry.

ii) an electrothermal atomiser, where a drop of the liquid sample is placed into a graphite tube, which is electrically heated. In this book, Chapter 6 is dedicated to electrothermal atomisation – atomic absorption spectrometry.

Some commercial instruments have now pre-aligned flame and furnace atomizers, making it feasible to select either of them simply by software commands.

Gaseous and volatilised analytes can also be easily determined by AAS using either flame or electrothermal atomisation. For this approach, the determination of several elements by formation of covalent volatile hydrides (e.g. arsenic, selenium) and cold vapour generation (mercury and cadmium) will be considered in Chapter 5. Moreover, in the last sections of Chapter 7 the coupling of gas chromatography to such atomisers is explained.

In addition to flames and electrothermal heated cells, other atomisers have been investigated for atomic absorption spectrometric measurements. For example, a jet-assisted glow discharge has been commercialised for the direct analysis of solid materials. The source can be used for bulk analysis of homogeneous solids and for depth profiling of layered materials. The atomisation cell is a vacuum chamber with two electrodes (the sample acts as the cathode). When the sample is inserted into position, the chamber is quickly evacuated and argon gas flows through the chamber. An electrical current ionises the argon, which bombards the cathode surface causing sample sputtering. The resultant sputtering is an efficient means of atomisation. The free atoms are entrained into a cohesive atom cloud into a space along the axis of the discharge chamber where absorbed light (from an external lamp) can be measured.

3.4. WAVELENGTH SELECTORS

Wavelength selection in UV-Vis spectrochemical instruments can be based on filters, spatial dispersion of wavelengths or interferometry. Wavelength selectors, which spatially disperse the spectral components of an optical beam, are the most common. Basically, they consist of an entrance slit (which defines the area of the source of radiation that is viewed), at least one dispersive element, which can be a prism or a grating, an image transfer system (mirrors and lenses), and a focal plane with one or several exit slits.

Most wavelength selectors used in AAS are monochromators. In such dispersive wavelength selectors, an exit slit about the same size as the entrance slit is used to isolate a small band of the spatially dispersed wavelengths that strike the focal plane. One wavelength band at a time is isolated and different wavelength bands can be selected sequentially by rotating the dispersive element to bring the new band into the proper orientation so that it will pass through the exit slit. The *linear dispersion*, D_l, specifies how far apart in distance two close wavelengths are separated in the focal plane. D_l is conveniently expressed in $mm.nm^{-1}$ units. Most often it is the *reciprocal linear dispersion*, R_D that is specified for a monochromator. R_D represents the number of wavelength intervals (e.g. nm) contained in each interval of distance (e.g. mm) along the focal plane.

The *spectral band-width* (SBW) or spectral band-pass is the half-width of the wavelength distribution passed by the exit slit. Except at small slit widths, where aberrations and diffraction effects must be considered, the SBW is controlled by the monochromator dispersive element and the slit width. The spectral band-width affects the spectral isolation (purity) of the analytical line and the required SBW is normally dictated by the nearest line adjacent to the analytical one in the spectrum.

Figure 3.6 shows the configuration of four types of monochromators used in commercial AAS instruments. Figure 3.6a shows schematically a grating monochromator based on the Czerny-Turner configuration. Incident radiation passes through the entrance slit and strikes a parabolic mirror (collimating mirror). The entrance slit, at the focal point of the collimating mirror, acts as a point source for the collimating mirror, which produces parallel radiation for the dispersing element (a grating). The grating spatially disperses the spectral components of the incident radiation. Collimated rays of the diffracted radiation strike a parabolic focusing mirror, which focuses dispersed radiation onto a focal plane. The exit slit, placed in this focal plane, isolates a particular narrow wavelength interval.

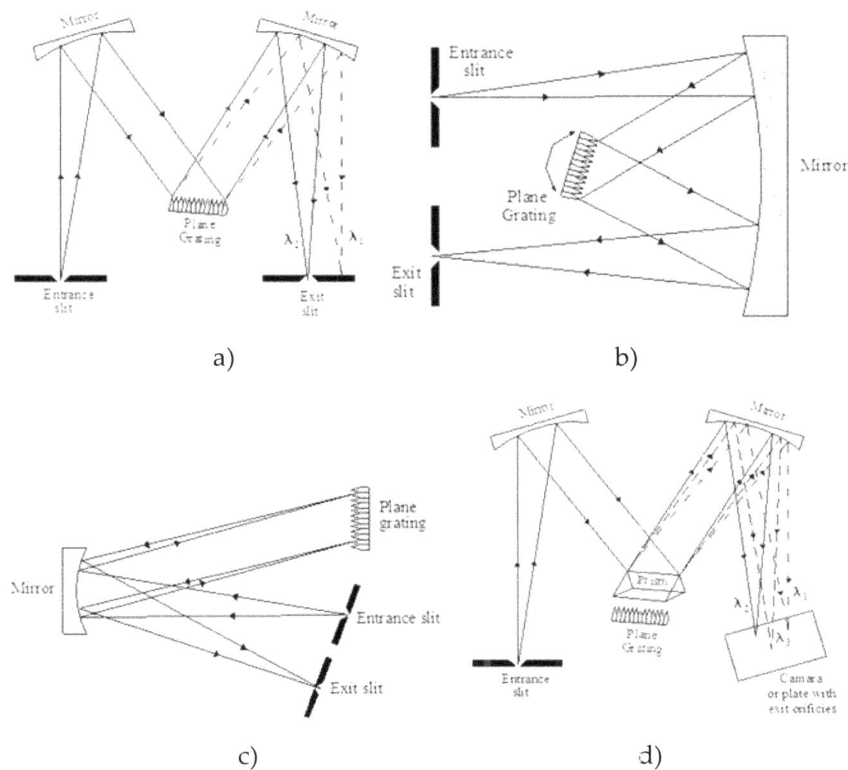

Figure 3.6. Dispersive wavelength selectors.

(a) *Czerny turner.* (b) *Ebert.* (c) *Littrow.* (d) *Echelle.*

The Ebert type (Figure 3.6b) is very similar to the Czerny-Turner except that one large mirror serves as both collimator and focusing mirror. The Czerny-Turner system has the advantage that focusing two mirrors is usually easier than focusing the single mirror of the Ebert monochromator. Also, two smaller mirrors are less expensive than one large mirror. Figure 3.6c shows the Littrow configuration. As can be seen, the Littrow configuration is a specific geometry in which the light of a specific wavelength diffracted from a grating, into a given diffraction order, travels back along the direction of the incident light (the deviation angle is close to zero). In some cases, a beam splitter is used to direct the diffracted radiation towards a more convenient focal plane in order to avoid to having the entrance and exit slits too close.

Finally, Figure 3.6d shows an echelle configuration. Originally invented for astronomy, echelle optics are now widely used in atomic spectrometry. It is a very compact optical design, incorporating usually a coarse grating (e.g. 120 grooves/mm) and a prism as a cross disperser, in a Czerny-Turner configuration. There are many designs of echelle optics, most using prisms but some using a grating as the cross disperser, although the basic configuration is similar.

The prism in the echelle monochromator serves as an order sorter. The coarse grating creates low resolution and different diffraction orders are dispersed in the same direction. Then the prism disperses the spectrum perpendicular to the diffraction direction, giving rise to enhanced resolution in a two-dimensional spectrum (the diffraction orders are separated by the prism). Therefore, instead of having a spectrum with intensity of every line against wavelength (i.e. one dimension), the echelle configuration creates a two dimensional spectrum with wavelengths across a horizontal line and a series of lines (order of the spectra) going down (like the text in a page of a book). In recent years, echelle optics is being incorporated into commercial AAS instruments for multielement analysis in combination with an auto-aligning turret (e.g. with eight HCLs). On the other hand, the advantages of using an echelle configuration in combination with a continuous primary source are obvious, since the details of the spectral background close to the line can be measured with a wider exit slit on a linear multichannel detector. Moreover, the use of such a two-dimensional multichannel detector allows for "true" simultaneous multielemental analysis.

3.5. DETECTORS

The intensity of the light passing the exit slit of a wavelength selector must be quantified with a proper transducer (see Figure 3.1). In this section the most commonly photon detectors will be briefly discussed.

The classical *vacuum phototube* consists of a photosensitive cathode and an anode in an evacuated glass or quartz enclosure. A high voltage is set between the two electrodes. Photo-irradiation of the cathode causes photoelectrons to be emitted and attracted to the anode, causing current to flow: a flow, which can be amplified and measured. Phototubes are used to detect moderate light levels.

In atomic spectrometry, however, the related and more sensitive photomultiplier tubes (PMTs) are preferred to vacuum phototubes. A *photomultiplier tube* can be considered as a vacuum phototube with additional amplification by electron multiplication. It consists of a photocathode, a series of dynodes on which secondary-electron multiplication occurs, and an anode. Photons strike the photocathode, which emits electrons due to the photoelectric effect. Instead of collecting these few first electrons (there should not be a lot, since the primary use for PMTs is in low intensity applications) at an anode, like in the phototubes, the electrons are accelerated towards a series of additional electrodes, called "dynodes", each at a more positive potential (50 to 90 V) than the one before it. Additional electrons are generated at each dynode. This cascading effect creates 10^5 to 10^7 electrons for each original photon hitting the photocathode. This amplified signal is finally collected at the anode where its intensity can be measured.

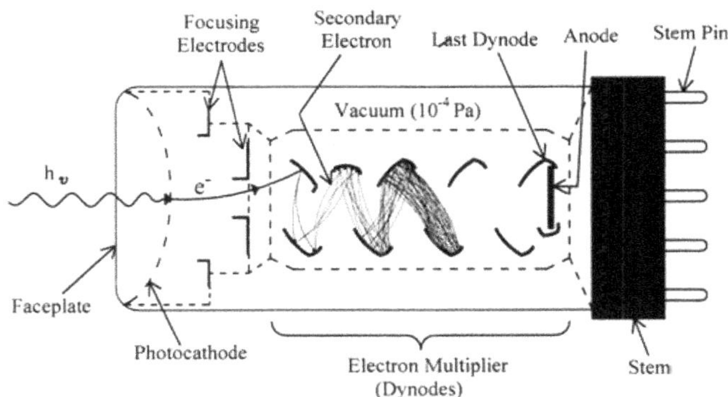

Figure 3.7. Diagram of a photomultiplier tube.

Figure 3.7 shows a diagram of a PMT. According to the desired characteristics in terms of response time, gain, etc, different types of photocathode and dynode structures have been developed. The PMT generally has the photocathode in either a side-on or a head-on configuration. The side-on type receives incident light through the side of the glass bulb, while the head-on (or end-on) type receives light through the end of the glass bulb. Figure 3.8 shows the external appearance of both configurations. The head-on type has a semitransparent photocathode (transmission mode photocathode) deposited upon the inner surface of the entrance window. Most side-on types employ an opaque (reflection mode photocathode) and a circular-cage structure electron multiplier.

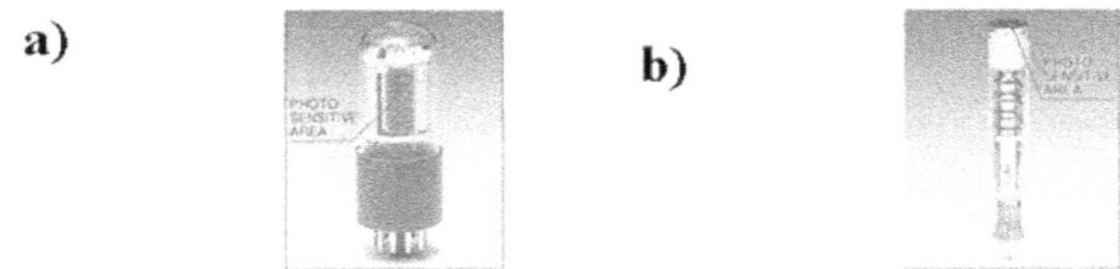

Figure 3.8. External appearance of photomultiplier tubes. Reproduced with permission of Hamamatsu. (a) *Side-on type.* (b) *Head-on type.*

The key properties of the PMT are numerous: large wavelength coverage, large dynamic range, high amplification gain, and low noise. However, the PMT is a single detector, and for this reason there is a current trend towards the replacement of PMTs by multichannel detectors. These consist of an assembly of many individual, small adjacent detectors called pixels. Charge transfer device (CTD) detectors are currently used, and are based on CCD or charge-injection device technology (CID). An assembly of linear arrays or a two-dimensional assembly can be used as the detector at the exit slit of the wavelength selector (in this case, the aperture of the exit slit is much larger than when a PMT is used).

3.6. BACKGROUND CORRECTORS

Contributions to the background of a measurement in AAS can arise from spectral interferences due to a spectral line of another element within the band-pass of the wavelength selector (such possibilities are rather uncommon in AAS; besides, such spectral interferences are now well-characterised), absorption by molecular species originated from the sample, and light scattering from solid or liquid particles present in the atomizer.

To obtain the accurate absorbance due to the analyte it is necessary to subtract the background from the total absorbance measured in the spectrometer. Thus, some commercially available strategies for this purpose will be discussed below. At this point, it is important to keep in mind that the ideal for background correction should be to measure a true blank solution.

3.6.1. Deuterium background corrector.

A common, comparatively inexpensive, background correction approach is based on the used of a continuum source. A deuterium lamp is the radiation source used to correct for background absorption and consists of a deuterium filled discharge lamp, which emits a continuum spectrum from 180 nm to about 400 nm, with a maximum near 250 nm. This is the spectral region where most analytical atomic absorption lines occur and where the effects of background absorption are most pronounced.

The deuterium lamp consists of a heated, electron-emitting cathode, a metallic anode, with a restrictive aperture between the two. A discharge current of several hundred milliamperes excites the deuterium gas emitting a continuum. The discharge is forced to pass through the small aperture, forming a defined area of rather high light emission. A suitable window allows light transmission to the spectrometer's optical system.

Figure 3.9a shows the optical configuration of a single beam spectrometer with deuterium background correction. In this configuration, the HCL and the deuterium lamp are sequentially pulsed on and off. When the HCL is on and the D_2 lamp is off, the total absorbance at the analytical wavelength is measured. This total absorbance comprises the absorbance from the analyte plus the absorbance from the background at that particular analytical line. When the HCL is off and the D_2 is on, the D_2 continuum energy fills the slit (see Figure 3.9b), and so the average of the background absorbance across the whole slit is measured. Since the spectral band-pass of the wavelength selector is usually large (e.g. 0.2 nm) compared to the analyte atomic line half-intensity width (less than 0.005 nm), the absorbance of the analyte is negligible compared to that of the average background when the deuterium lamp is used. Thus, the AAS true signal is calculated by subtracting the average of the background absorbance (deuterium lamp) from the total absorbance (measured with the HCL).

It is important that both the deuterium source and the hollow cathode lamp are well aligned to follow the same optical path. If they are not, then the two measurements may not be made on the same atom population and significant errors may occur. Moreover, to obtain successful background correction the intensity of the deuterium lamp must be matched to that of the hollow cathode lamp. For example, an approach to follow if the light from the continuum source is too intense consists of reducing the spectral bandwidth (the energy measured from the continuum source increases with the square of the spectral bandwidth, while the radiant energy measured of the atomic spectral line from the HCL increases linearly with the spectral bandwidth). Conversely, the spectral band-width should be increased if the HCL intensity is too high for the deuterium lamp.

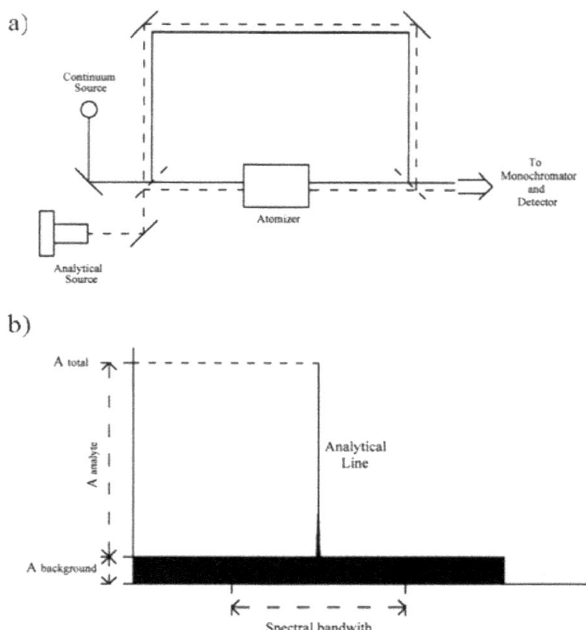

Figure 3.9. Deuterium background corrector.
(a) Schematic of a single-beam AAS spectrometer with deuterium background correction.
(b) Basis of the correction. A_{total}: total absorbance at the analytical wavelength. $A_{analyte}$: absorbance due to the analyte. $A_{background}$: absorbance at the analytical wavelength produced by the background.

Unfortunately, the deuterium background corrector cannot accurately correct for either high absorbances or for a structured molecular background (since the average of the background absorbance measurements may not be representative of the actual background at the analytical line).

3.6.2. Zeeman correction

The splitting by a magnetic field of the spectral lines of an atom, as well as the polarization of these lines, constitutes the basis of background correction by the Zeeman effect.

Principles of Zeeman Effect

The quantum states of an atom undergo drastic changes when that atom is placed in a magnetic field; energy states that were "degenerate" may separate from each other and so spectral lines may split into three or more components. Transitions between these new states are given by the usual selection rule:

$\Delta M_j = 0, !1$

where M_j is the magnetic quantum number.

The $\Delta M_j = 0$ components (called the π components) have their electric vectors linearly polarized parallel to the magnetic field, while the $\Delta M_j = !1$ components (called the $\sigma!$ components) have their electric vectors linearly polarized perpendicular to the magnetic field. The $\sigma!$ components are circularly polarized about the lines of force, also, with the σ^+ y σ^- component vectors rotating in opposite directions.

The "normal" Zeeman effect (called normal because it was explicable by classical physics) arises either when both states π and σ are split by an equal amount of energy. A triplet is then obtained with a central π component and two σ^- and σ^+ components (see Figure 3.10). In this case, the component separation is directly proportional to the magnetic field. The "anomalous" Zeeman effect is more complex and leads to the formation of a multiplet the separation of which is a more complicated function of the magnetic field.

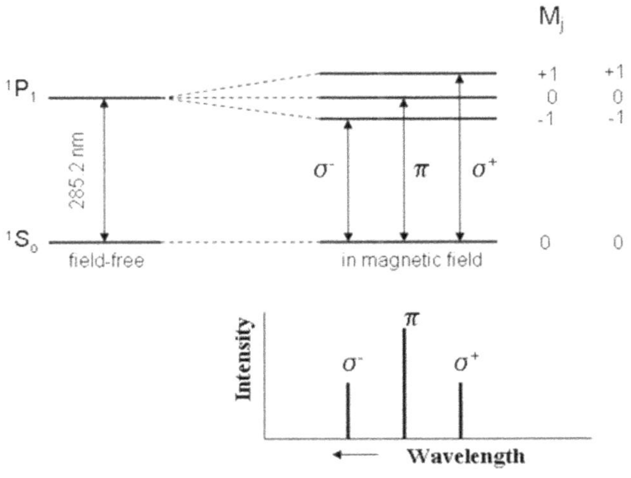

Figure 3.10. Normal Zeeman effect in the 285.2 nm magnesium line.

Figure 3.11 shows a diagram of the optical system of a commercialised AAS with an AC transverse (magnetic field applied across the light path) Zeeman corrector. As can be seen, a modulated electromagnet is located around the atomizer (a graphite furnace). A polarizer is positioned between the atomisation cell and the wavelength selector to remove the π component of the transmitted radiation. When the field is off, the total absorbance is measured (analyte and background), while when the field is on only the background is measured. Therefore, the background measurement can be made at the exact wavelength when the magnetic field is applied.

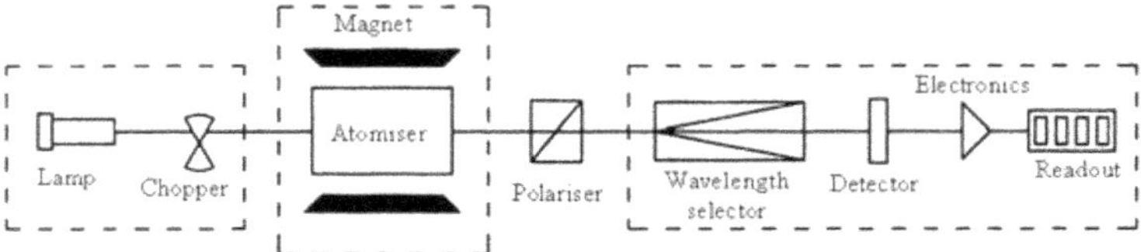

Figure 3.11. Diagram of the configuration of an AAS with Zeeman background correction.

In some configurations the polarizer is located prior the atomisation cell and a permanent magnet is used. By alternately polarizing the incident light parallel and perpendicular to the magnetic field, the background absorption is measured first and then the background plus sample absorption. This is because the σ± components have been shifted away from the sample atoms' resonance line and the remaining π components of the resonance line absorb only parallel polarized light, while the background equally absorbs both polarizations of light.

All Zeeman systems are optically single beam spectrometers and require only a single energy source. Although the magnetic field could be applied to the lamp (direct Zeeman effect), most commercially available AAS systems apply the magnetic field to the atomiser (inverse Zeeman effect). The magnetic field is characterised by its mode (transverse or longitudinal) and its frequency. DC Zeeman background correctors use a permanent magnet and a rotating or vibrating polarizer to separate the combined and "background only" signals. AC systems use an electromagnet, and measure the combined and "background only" signals by switching the magnetic field on and off. An important difference between longitudinal (magnetic field applied along the light path) and transverse AC Zeeman systems is that transverse systems use a polarizer, while the longitudinal configuration does not require the polarizer.

Advantages of Zeeman background correctors include that they enable background signals to be measured at exactly the analytical wavelength. Therefore a structured molecular background and spectral interferences are easily corrected for. Besides, this strategy corrects for high levels of background absorption. A disadvantage of the Zeeman background corrector, however, is "calibration roll-over". Calibration roll-over occurs when the σ components of the analyte are not sufficiently shifted from the analyte line so that an overlap occurs at the absorption line. Roll-over can cause two different analyte concentrations to have the same absorbance (to prevent such roll-over problems occurring in practice, the maximum permissible absorbance should be known). Another possible disadvantage of the Zeeman effect is a loss of sensitivity for some elements.

3.6.3. Smith-Hieftje correction

In the Smith-Hieftje background correction method the HCL is pulsed alternatively at low and then at very high current (see Figure 3.12). The use of high currents during short periods of time does not give rise to melting problems (and therefore shorter lamp lifetimes) of the HCL since the average current is low.

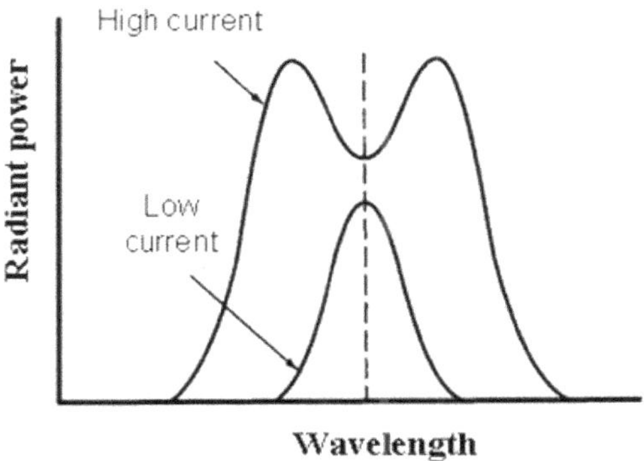

Figure 3.12. Emission line profiles in hollow cathode lamps operated at high and low currents.

During the low current part of the cycle (see Figure 3.12) the line is narrow and the analyte and background are measured together. However, during the brief high current pulse, the lamp emission lines broaden as a large cloud of atoms is formed in front of the lamp cathode. This cloud essentially prevents resonant radiation from reaching the analyte in the atomiser due to HCL emission self-reversal (see Chapter 1). Thus, in this part of the cycle mainly the background is probed.

This method is simple, does not require additional components and can be used with any type of atomizer.

Chapter Four

FLAME ATOMIC ABSORPTION SPECTROMETRY

4.1. INTRODUCTION

As stated in previous chapters, for AAS measurements the analyte must be converted into free gaseous atoms, usually by the application of heat in the atomizer (e.g. using a flame or an electrically heated furnace). Analytical atomic spectroscopy based on the use of a flame as atomizer was first developed in the middle of the 20th century, being at that time flame atomic emission spectrometry (FAES) the precursor of flame atomic absorption spectrometry (FAAS). In 1964, with FAES and FAAS already enjoying a strong development, Winefordner in the USA and West in the United Kingdom demonstrated that atomic fluorescence in flames could be also used for analytical purposes.

Other atomizers offering higher analytical sensitivity than flame have been developed with high success (e.g. graphite furnace/electrothermal atomization in AAS). However, flame-based atomic instruments are still widely employed, particularly for routine applications in inorganic elemental analysis. In Chapter 2 it was emphasized that the continued use of flame-based atomic spectrometric instruments is mainly due to the facts that they are very robust and comparatively cheap, their analytical methods are well established and validated, and the analyses are rapid. At present, flame-based atomic techniques can still be considered as a cost-effective choice whenever the expected concentrations of the analyte are in the $mg.L^{-1}$ level .

It should be stressed again here that flame methods based on atomic emission, atomic absorption and atomic fluorescence are usually considered as complementary (see Chapter 2), but FAAS is by far the most commonly used, providing reliable results when used even by people with minimal training.

4.2. THE ATOMIZER UNIT IN FLAME ATOMIC ABSORPTION SPECTROMETRY

Although methods have been developed in which solid samples are introduced directly or as suspensions into a flame for AAS, the analysis of liquid samples and dissolved solids is much more common (FAAS can be also used for volatile or volatilized analytes and Chapter 5 addresses this topic). As can be seen in Figure 4.1, a liquid sample containing the analyte, M, once in the atomizer undergoes several stages to obtain the desired maximum number of analyte atoms in the fundamental required state. Typically, in FAAS the liquid sample, in the first step, is converted into a fine spray or mist (this step is called nebulization). Then, the spray reaches the atomizer (flame) where a series of processes, including desolvation, volatilization and dissociation, take place to produce gaseous free atoms. As also shown in Figure 4.1, once the gaseous free atoms of M are formed, they can suffer other processes in the flame, such as ionization, excitation and chemical reactions. We will now explain each process in more detail:

Desolvation. Evaporation of the solvent is the first step to occur after nebulization, leaving a dry aerosol of molten or solid particles. Water tends to lower the flame temperature; however, desolvation efficiencies can be quite high when larger droplets are removed prior to entering the flame (in the nebulization chamber). Organic solvents evaporate more rapidly than water and the desolvation can be accelerated by the heat of combustion of the organic vapor.

Volatilization and dissociation. The solid or molten particle remaining after desolvation must be vaporized to obtain gaseous molecules, which are then dissociated to produce the sought after free atoms. Thus, incomplete volatilization leads eventually to a loss of analyte free atoms and so to a reduction in the analytical signal. It can also be detrimental by causing nonlinearity in analytical curves and continuous background emission from incandescent particles and light scattering.

Ionization, excitation and chemical reactions. In the vapor phase of the flame the analyte can exist as free atoms, molecules or ions. The formation of molecular species and ions reduces the concentration of free atoms and thus degrades the detection limit. Molecular species, such as MO, MH or MOH, can be formed by reactions of analyte atoms with gaseous flame constituents or with volatilized species from the sample (high temperatures and a reducing environment tend to reduce the formation of refractory oxides). As explained in Chapter 2, ionic species are also formed by the loss of an electron from an atom. Therefore, for analytes with low ionization potentials the use of a low-temperature flame and a spectroscopic buffer (readily producing electrons to retrograde the equilibrium) are advisable.

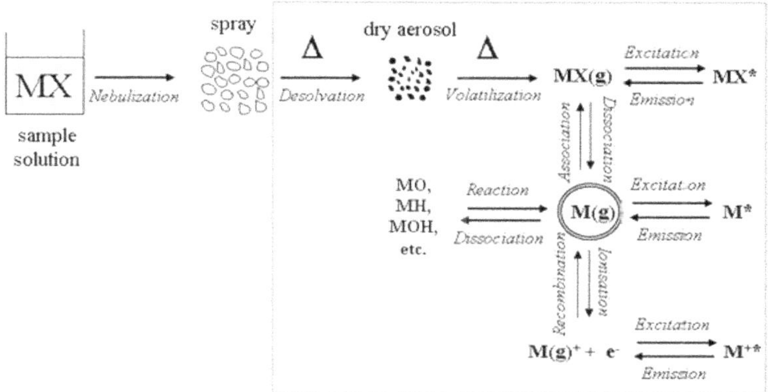

Figure 4.1. Diagram showing the steps to obtain fundamental analyte (M) atoms from a liquid or dissolved sample containing the salt MX. (*): excited state.

Finally, as shown in Figure 4.1, in some cases the flame can be efficient enough to excite gaseous atoms and generate spontaneous emission of radiation characteristic of the analyte.

4.2.1. Nebulizer, nebulization chamber and burner.

The most commonly used nebulization systems are pneumatic devices in which a jet of compressed gas (the nebulization gas) aspirates and nebulizes the solution. Figure 4.2 shows a picture and a diagram of a typical pneumatic concentric nebulizer used in FAAS. In such a device, the gas flows through a small opening that concentrically surrounds the capillary tube, causing a reduced pressure at the tip (Venturi effect) and thus a continuous suction of the sample solution from its container (giving rise to steady-state analytical signals). In most cases, the aspiration rate is proportional to the pressure drop along the capillary and inversely proportional to the viscosity of the solution. The sample solution, drawn up the capillary tube, encounters at the exit the high velocity of the nebulizing gas, which causes the formation of droplets of various sizes. Pneumatic nebulizers produce droplet diameters typically in the range 1 to 50 μm.

Nebulizers are commonly made of stainless steel (preferred for general applications with acid concentrations less than 5%). However, for maximum chemical resistance there are also available platinum alloy nebulizers (not suitable for use with aqua regia or hydrofluoric acid) consisting of a platinum alloy (e.g. Pt-Ir) needle assembly and a venturi made of materials such as tantalum or poly(etheretherketone) (PEEK). The connection between the nebulizer and the liquid sample is usually a piece of plastic capillary tubing (about 15 cm long). Of course, it has to be always ensured that this plastic capillary tubing is correctly fitted to the nebulizer capillary; any leakage of air, tight bends, or kinks will cause unsteady, non-reproducible readings.

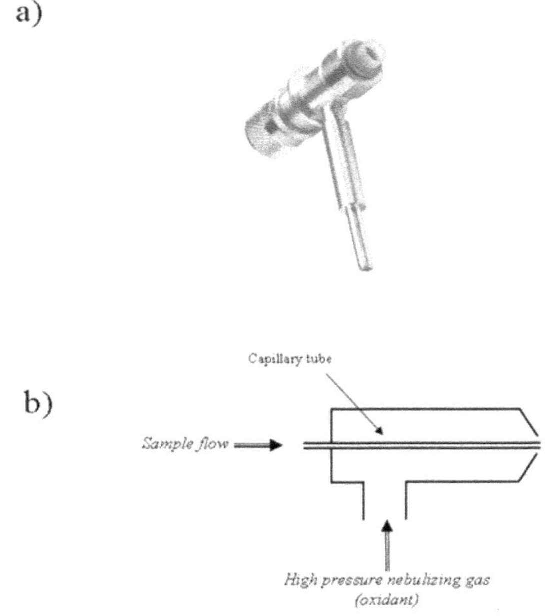

Figure 4.2. Nebulizer used in FAAS.

a) Photograph of a typical nebulizer. b) Basic schematics of a venturi-based nebulizer.

> At times the plastic capillary tubing can become clogged and it will be necessary to cut off the clogged section or fit a new piece of capillary tubing. The nebulizer capillary can also become clogged. If that occurs, the flame should be turned off and the nebulizer dismantled and placed in an ultrasonic cleaner containing liquid soap solution. If the ultrasonic bath fails to clear the blockage, a blur-free wire should be carefully passed through the nebulizer and then the ultrasonic cleaning procedure repeated.

Two types of burners have been used in flame spectroscopy: **total consumption burners** and **premix chamber burners**. In the **total consumption burner,** a concentric nebulizer without a spray chamber is used. A third concentric orifice transports the fuel gas around the nebulizer tip so that the nebulizer is an integral part of the burner (see Figure 4.3). The flame is supported immediately above the nebulizer tip. Thus, when using pneumatic concentric nebulizer-burners the entire aspirated sample is directed into the flame, without any droplet size selection, allowing the introduction of relatively large amounts of sample into the flame. However, these burners are noisy (both from the electronic and audible standpoints) and other disadvantages include a relatively short pathlength through the flame and problems with tip clogging.

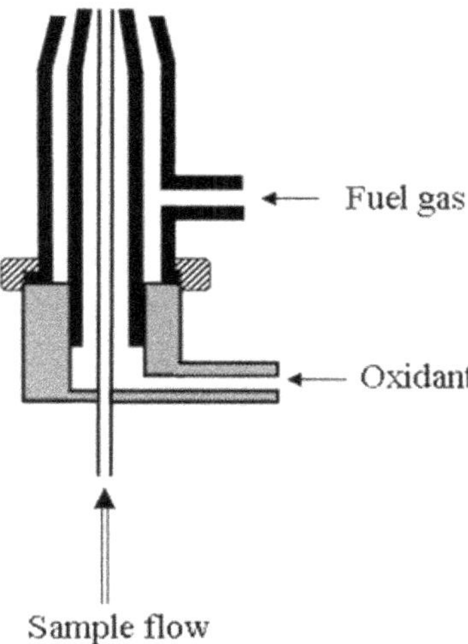

Figure 4.3. Basic schematics of a total consumption burner.

Nowadays, the use of total consumption burners is almost neglected in FAAS. Commercial instruments today use commonly a **premix chamber** (see Figure 4.4). In the premix chamber, large droplets of the nebulized sample solution condense and drain out of the chamber ("nebulization chamber"), while the remaining fine droplets mix with the fuel and oxidant gases before they enter the burner head. Devices that remove large droplets usually consist of a paddle (e.g. flow spoiler) or an impact bead. Typically about 90% of the sample droplets condense, leaving only 10% of the sample to enter the flame. In flame atomisers, the nebulization gas is usually the oxidant (e.g. air), and the fuel is brought into the nebulization chamber through a separate port. The mixed oxidant, fuel and fine sample mist are then carried to the burner head, which supports the flame. Although a large portion of the aspirated sample is lost in the drain of the premix chamber, the "atomization efficiency" of that portion of the sample entering the flame is greater than in total consumption burners, because the droplets are much finer. Also, the pathlength is longer. In addition, combustion with the premix burners are very quiet (less audibly noisy signals).

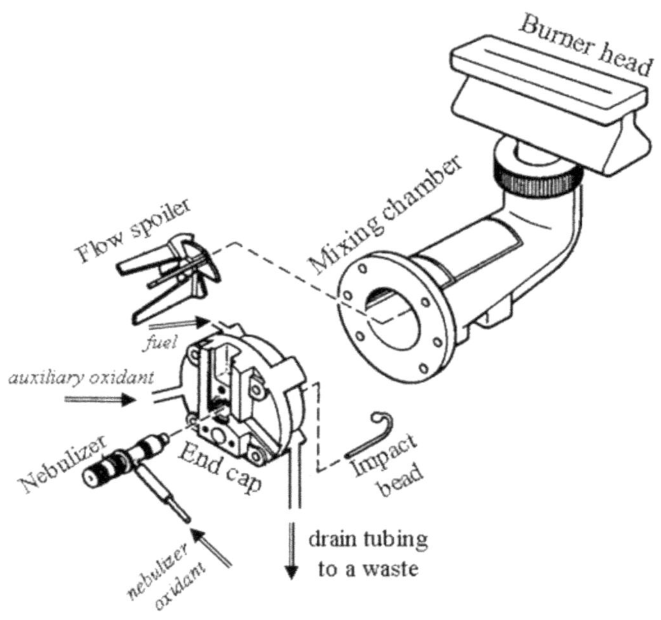

Figure 4.4. Example of a nebulizer pre-mix chamber (it can be used with either a flow spoiler or an impact bead) and burner head.

A drain vessel must be used to gather the effluent from the premix chamber drain. The drain vessel should be checked regularly and replaced when necessary. Therefore, the vessel should not be stored in an enclosed storage area. Rather, it should be sited in plain sight of the operator, usually at the front of the instrument or on an open shelf underneath the instrument table.

In Figure 4.4 a typical burner head of a premix burner is also shown. As can be seen, the slot, which is usually aligned with the instrument's light path, is long and thin (typically several cm long and tenths of mm wide). Some companies also market specially designed burners (e.g. a three-slot burner head), which do not clog easily even when concentrated solutions are used. Burner heads are made completely from titanium or of special corrosion resistant alloys.

> During aspiration of certain solutions, carbon and/or salt deposits can build up on the burner causing changes in the fuel/oxidant ratio and flame profile. To minimize the accumulation of salts, a dilute solution of acid may be aspirated between samples. However, if salts continue to build up, the flame should be turned off and an appropriate cleaning strip should be inserted in the burner slot and moved back and forth through the slot (care should be taken to not use sharp objects such as razors to clean the burner as they can damage the slot and form areas where salt and carbon can accumulate at an accelerated rate). If such cleaning proves to be inadequate the burner should be removed and soaked in warm soapy water, or in dilute acid, or in an ultrasonic bath with non-ionic detergent.

4.2.2. Flame

Common AAS flames are today obtained with a premix burner where fuel and oxidant are mixed in a previous chamber before combustion in the burner (see Figure 4.4). These "premix" flames are very stable and show a structure consisting of cones (see Figure 4.5). The appearance and the size of these regions depends on the ratio "fuel/oxidant" and also on the type of fuel and oxidant used. In the **inner zone** (primary combustion zone) the fuel/oxidant is ignited and combustion reactions proceed rapidly to completion or as far as the available oxidant allows. The **interconal zone** (which corresponds to the hottest part of the flame) lies immediately above. In that region, local thermal equilibrium exists and is the region in which most analytical measurements are carried out. This interzonal region is externally bounded by the **outer zone** (secondary combustion zone), which is much more diffuse and less well defined than the primary combustion zone. This external region is in contact with the atmosphere, so, oxygen and nitrogen from the surrounding air can penetrate and cause additional reactions here.

The part of the flame to be used in a given analysis depends on the analyte (generally the flame position is adjusted to yield a maximum absorbance reading). It is important to note that though the temperature of the flame is of great significance, the oxidizing or reducing characteristics of the flame are also of paramount importance. For example, Figure 4.6 shows the observed absorbance of three elements as a function of distance (height) above the burner tip. For magnesium and silver, the initial rise in absorbance is a consequence of a longer exposure to heat, which leads to a greater concentration of atoms in the radiation path. However, the absorbance for magnesium reaches a maximum at about the center of the flame and then falls off with height as oxidation to magnesium oxide takes place (the effect is not seen with silver because this element is more resistant to oxidation). For chromium, which forms very stable oxides, the maximum absorbance lies immediately above the burner tip.

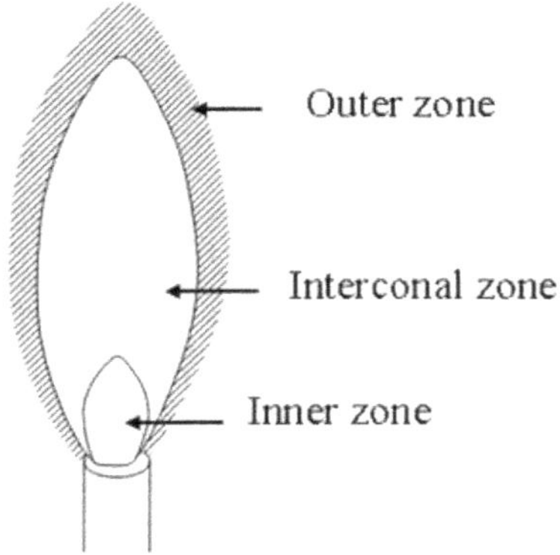

Figure 4.5. Typical flame regions.

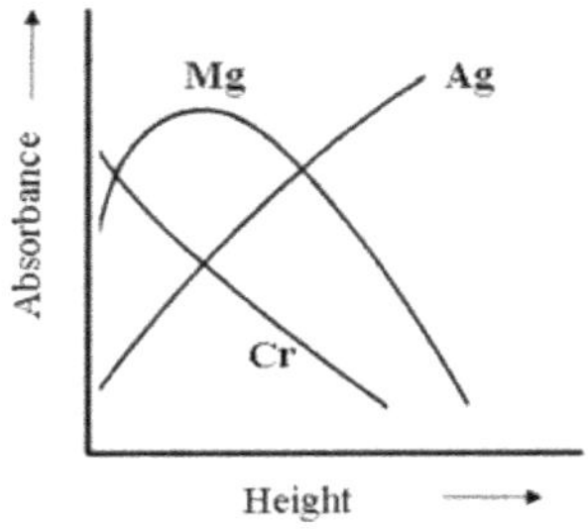

Figure 4.6. Illustrative flame absorbance profile for magnesium, silver and chromium, depending on height of observation of the atoms in the flame.

Several combinations of fuel and oxidant can be employed in FAAS methods. Table 4.1 lists the typical temperatures reached with most common flames together with their burning velocities. The burning velocity is a critical flame parameter, as flames achieve stability only within a certain region of gas flow rates. If the gas rise velocity does not exceed the particular flame burning velocity, the flame will propagate inside the burner, resulting in a flashback condition (an eventual explosion is most likely). As the gas flow is increased, the rise velocity soon exceeds the burning velocity at all points and the flame rises until it reaches a steady state point above the burner where the rise and burning velocities are just equal. This is the region of a stable flame. At yet higher flow rates, the flame would continue to rise and would finally blow off. Flames that burn with air as the oxidant can be premixed, because of their relatively low burning velocities. The nitrous oxide-acetylene flame can be premixed with some care. However, the more rarely used mixtures of oxygen and acetylene require a great deal of care to burn safely after premixing. Of course, oxygen and hydrogen may also form explosive mixtures, but if diluted with argon or helium, they can be premixed.

Table 4.1

Characteristics of some analytical flames.

		Maximum burning	
Fuel	Oxidant	Temperature (^{o}C)	velocity ($cm.s^{-1}$)
Acetylene	Air	2100-2400	158-266
Acetylene	Nitrous oxide	2600-2900	285
Acetylene	Oxygen	3050-3150	1100-2480
Hydrogen	Air	2000-2100	300-440
Hydrogen	Oxygen	2550-2700	900-1400
Propane	Air	1700-1900	39-43
Propane	Oxygen	2700-2800	370-390

> A venting system is required to remove the combustion fumes and vapors from the flame atomic absorption instrument. This venting system will protect laboratory personnel from toxic vapors which may be produced by some samples, it will help to protect the instrument from corrosive vapors that may originate from the sample, and it will remove dissipated heat, which is produced by the flame or furnace.

Nowadays, the most widely used gas-combinations for FAAS are air-acetylene and nitrous oxide-acetylene. This latter high-temperature flame is very useful for those elements that tend to form heat-stable oxides in the air-acetylene flame (the "refractory elements"). However, it is not always required and may even be detrimental for many elements, because its higher temperature will cause undesired ionization of gaseous atoms. Depending on the burner head, the flame is generally 5 to 10 cm long (because of the somewhat higher burning velocity, for the hot nitrous oxide/acetylene flame, specially designed burner heads with a narrow slot are required to prevent flashback of the flame).

Figure 4.7 shows a periodic table indicating which of the two flames are recommended for each particular element. Below are collected characteristics, peculiarities and precautions to use with the three more common gases used in FAAS: air, acetylene and nitrous oxide.

H																	He
670.8 Li 0.7	234.9 Be 0.7											249.7 B 0.7	C	N	O	F	Ne
589.0 Na 0.2	285.2 Mg 0.7											309.3 Al 0.7	251.6 Si 0.2	213.6 P 0.2	S	Cl	Ar
766.5 K 0.7	422.7 Ca 0.7	391.2 Sc 0.2	364.3 Ti 0.2	318.4 V 0.7	357.9 Cr 0.7	279.5 Mn 0.2	248.3 Fe 0.2	240.7 Co 0.2	232.0 Ni 0.2	324.8 Cu 0.7	213.9 Zn 0.7	287.4 Ga 0.7	265.1 Ge 0.2	193.7 As 0.7	196.0 Se 2.0	Br	Kr
780.0 Rb 0.7	460.7 Sr 0.2	410.2 Y 0.2	360.1 Zr 0.2	334.4 Nb 0.2	313.3 Mo 0.7	261.4 Tc 0.2	349.9 Ru 0.2	343.5 Rh 0.2	244.8 Pd 0.2	328.1 Ag 0.7	228.8 Cd 0.7	303.9 In 0.7	286.3 Sn 0.7	217.6 Sb 0.2	214.3 Te 0.2	I	Xe
852.1 Cs 0.7	553.6 Ba 0.2	550.0 La 0.2	286.6 Hf 0.2	271.5 Ta 0.2	255.1 W 0.2	346.0 Re 0.2	290.9 Os 0.2	264.0 Ir 0.2	265.9 Pt 0.7	242.8 Au 0.7	253.7 Hg 0.7	276.8 Tl 0.7	283.3 Pb 0.7	223.1 Bi 0.2	Po	At	Rn
Fr	Ra	Ac															

Ce	495.1 Pr 0.2	492.4 Nd 0.2	Pm	429.7 Sm 0.2	459.4 Eu 0.2	368.4 Gd 0.2	432.6 Tb 0.2	404.6 Dy 0.2	410.4 Ho 0.2	400.8 Er 0.2	371.8 Tm 0.2	398.8 Yb 0.2	336.0 Lu 0.7
Th	Pa	351.5 U 0.2	Np	Pu	Am	Cm	Bk	Cf	Es	Fm	Md	No	Lw

Figure 4.7. Periodic table with elements that can be analyzed by FAAS.

Compressed air. Air compressors are commonly used. It is strongly recommended to have a water and oil trap or filter between the compressor and the instrument. Air compressors are generally noisy and whenever possible it is advisable to locate them in a suitable ventilated area at some distance from laboratory workers. Cylinders of compressed air can be also used. However, they are recommended only as a short-term solution: a premix burner-nebulizer system uses about 20 $L.min^{-1}$ of air and, therefore, a cylinder will last only a few hours (so, unless a FAAS instrument is used only sporadically, changing cylinders becomes a labour taking risk with the added disadvantage of being expensive in the long run).

Acetylene. Suitable acetylene has typically a minimum purity specification of 99.6%. Acetylene is normally supplied dissolved in acetone, and a small amount of acetone carry-over with the acetylene is acceptable (for better conditions, it may be desirable to have an acetylene filter between the acetylene tank and the instrument gas control system to remove particulates and acetone droplets from acetylene, thereby protecting the gas controls). However, as tank pressure falls, the relative amount of acetone entering the gas stream increases and can give rise to erratic results, as well as damaging instrument connecting pipes. For this reason, acetylene tanks should be replaced when the cylinder pressure drops to about 600 kPa. Furthermore, acetylene tanks should always be stored and operated in a vertical position to prevent liquid acetone from reaching the cylinder valve. As an important precaution, acetylene line pressure from the cylinder to the instrument should never be allowed to exceed 100 kPa. At higher pressures, acetylene can spontaneously decompose or even explode.

Nitrous oxide. The use of cylinders of nitrous oxide (99.0% minimum purity) requires a number of accessories and precautions. Nitrous oxide is supplied in the liquid state. Since the nitrous oxide is in liquid form, the pressure gauge does not give a true indication of how much nitrous oxide remains in the cylinder until the pressure starts to fall rapidly as the residual gas is drawn off. When nitrous oxide is rapidly removed from the cylinder, the expanding gas causes cooling of the cylinder pressure regulator and the regulator diaphragm sometimes freezes. This can create erratic flame conditions or even a flashback. It is therefore advisable to heat the regulator. An important precaution is that all lines carrying nitrous oxide should be free of grease, oil or other organic material, as spontaneous combustion could eventually take place.

4.2.3. Special sampling techniques

One of the ever existing aims in AAS has been the improvement of the achievable detection limits, which are primary limited by the usual sample introduction system (called "the *Achille's heel*" of analytical atomic spectrometry). As explained above, in the conventional nebulizer/burner systems only about 10% of the aspirated sample solution reaches the flame, while the other 90% goes to waste. Moreover, the transport of the sample to the flame is limited by the aspiration rate of the nebulizer. These two disadvantages were addressed in the late 1960s and early 1970s with different approaches including the so-called *sampling boat technique* and the *Delves cup*. Nowadays, such initial approaches have been displaced by ETAAS technique (see Chapter 6). However, they will be briefly described below because they are of interest to understand this problem and its modern solutions.

In the "sampling-boat technique" a small amount of sample is introduced into a tantalum boat, the solution is dried (for example, near the flame of an atomic absorption spectrometry or in a furnace) and then the boat is introduced into the flame of the spectrometer. The sample in the boat is in the flame rapidly and quantitatively atomized giving rise to a transient-signal. The temperature that can be achieved with this technique is only around 1200°C, but this temperature is suitable for some elements of toxicological and environmental interest, such as As, Cd, Pd, Hg, Se, etc.

A little later, Delves proposed a modification of that boat technique, giving rise to the "Delves cup", which enjoyed wide use for the determination of lead and cadmium in biological fluids. In this case, the long tantalum boat was replaced with a small round nickel cup, and an open tube was mounted over this cup to increase the sensitivity. The radiation beam passed through the tube and the analyte element was atomized into the tube. This tube increased the residence period of the atoms in the optical path.

4.3. FLAME ATOMIC ABSORPTION INSTRUMENTATION

Today, at least ten well-known companies manufacture atomic absorption spectrometers (see section 8.1 in Chapter 8). Most of these companies have several models of AAS instruments available for customer choice. In some cases, marketed AAS spectrometers can be used both with a flame and with an electrothermal atomizer, which are easily exchanged if needed; there are also instruments with true simultaneous operation of flame and graphite furnace, both atomizers permanently mounted and aligned for immediate use, available commercially.

Nowadays AAS manufacturers market an important variety of accessories for automated measurements or to enhance the applicability range and performance of the offered AAS instrumentation.

4.3.1. Flame atomic absorption spectrometers

There is today a wide choice of instruments for FAAS measurements, including single beam or double beam optics. For background correction the deuterium lamp (although other choices are available) is preferred in FAAS work.

Some of the FAAS instruments have modular parts that can be easily replaced, so providing a high degree of flexibility. In some cases they have complex and expensive optics for maximising performance, while more attention to other requirements is attended by others (e.g. to simplicity in order to make an AAS instrument portable for field applications).

Although we explained why AAS is considered a "single-element technique" in Chapter 2 (implying that aspiration of every sample has to be repeated for each element to be determined), nowadays fast-sequential AAS instruments combined with an automatic turret for up to eight hollow cathode lamps are commercially available, enabling several elements to be determined in rapid sequence in just a "single aspiration" of each sample (this capability also allows for internal standard corrections to compensate long term drift or sample preparation errors). Besides, it should also be mentioned here that a company markets an AAS instrument with a continuous source (a xenon short-arc lamp) and a high-resolution echelle spectometer with UV-sensitive CCD detector **allowing for multielemental analysis** and simultaneous background correction [**Becker-Ross *et al*** (2002), **Welz *et al*** (2003)].

Finally, it has to be noted that in the last few years particular attention has been given by most companies to secure flame safety. This goal is being achieved, for example, by monitoring critical components such as flame ignition, pressure in gas supplies and in the mixing chamber, drain status, automatic recognition of installed burner head to ensure proper re-adjustment of the correct gas settings for the type of flame used, etc,.

4.3.2. Accessories

Several manufacturers market several similar-use commercially available accessories for flame atomic absorption spectrometers with slight modifications, although a single manufacturer may market a specialised accessory. If the reader is interested, this particular information can be found through a quick search of the webpages of the AAS companies given in section 8.1 of Chapter 8.

To restrict this section, information about commercial versions of hydride or cold vapor generation can be found in Chapter 5 and information about accessories to carry out AAS measurements in flow systems is provided in Chapter 7.

4.3.2.1. Autosamplers. Many companies market computer-controlled fully automated autosamplers with *xyz* arm movement and full random access capability. They usually have a capacity of more than one hundred samples and several standards (in some designs unlimited sample capacity is claimed because in such cases it is possible to exchange sample racks during an auto-analysis cycle). Other features incorporated in some available autosamplers are, for example: variable probe arm speed settings for samples of different viscosities, elimination of cross-contamination by the use of a pump providing a continuous stream of clean rise solution, variable on-line dilution, automated standard additions, etc.

4.3.2.2. Atom concentrator tube or slotted tube atom trap. Several companies market a side-open quartz tube, with two slots along its length, which is positioned in the air/acetylene flame. The flame enters the tube through one of the slots and exits through the other and both ends of the tube. It is claimed that this system enhances the sensitivity by 2-3 times when compared to that of normal FAAS.

When a solution is aspirated, the atoms pass into the tube. The tube wall prevents the atoms from diffusing out of the optical path. This increases the residence time of the atoms in the light beam, thus increasing the absorbance signal. The tube also stops air diffusing into the flame, preventing oxides formation, a cause of poorer sensitivity.

4.3.2.3. High solids analyser. The analysis of samples with high dissolved solids content has always been a problem in FAAS, because of the tendency of dissolved salts to crystallize and cause blockages in the nebulizer tip or in the burner. A common way to avoid this problem is to dilute the sample until the concentration of dissolved solids is less than 2% before the final analysis; several manufacturers market automatic sample dilution systems. Also, a high solids analyser is marketed that allows for the analysis of solutions containing as much as 30% total dissolved solids without blockage of the nebulizer or the burner. The accessory consists of a switching valve, incorporating a sample loop of a predetermined volume (100 μL), which allows the introduction of a small volume of sample into a rinse liquid stream and passage of the sample-containing zone into the spray chamber with minimal diffusion. The aspiration rate of the sample is determined by the nebulizer uptake rate and, as expected, the discrete sample introduction produces a transient absorbance signal.

4.3.2.4. Flame microsampler. When only a small amount of sample is available, or where the burner could become clogged by continuous aspiration of high-solids for a long while (e.g. solutions of serum, sea-water, ore and metallurgical digests), a microsampler system (syringe 50-250 μL) for direct injection can be used.

4.3.2.5. Automatic burner rotation. One inherent problem with conventional AAS analysis is its comparatively narrow dynamic ranges (that is to say that samples with a wide analyte concentration range cannot always be directly measured in a single calibration). To overcome such problems two options are available: (i) to resort to an automatic on-line dilution, or (ii) to employ a system allowing for automatic burner rotation. Such a "burner rotation system" provides a monitored adjustment, controlled from the instrument PC, in order to have the burner slit accurately and reproducibly positioned in the light beam to obtain different "light path" values, L, and so optimum absorbance measurements (Abs = ε.L.C).

4.4. ANALYTICAL PERFORMANCE CHARACTERISTICS AND INTERFERENCES

Interferences in atomic spectrometric techniques have been classified in Chapter 2; here the extent of each type of interference as relevant in FAAS measurements only will be discussed.

4.4.1. Spectral interferences

Spectral interferences are usually strongly dependent on the spectral bandwidth of the monochromator. These interferences may arise from *overlapping concomitant atomic absorbance lines* (they are rare because the emission lines of hollow cathode lamps are very narrow; in any case, they can be avoided by choosing a different absorbance line) or by *molecular absorption or scattering*, e.g.:

a) from molecular absorption or scattering of source radiation by unvaporized solvent droplets or molecular species from the sample (they will cause a positive interference in AAS). A potential matrix interference due to absorption occurs, for example, in the determination of barium in alkaline earth mixtures. The barium wavelength used for AAS analyses lies in the center of a broad absorption band for molecular CaOH, thus resulting in interference by calcium in the barium determination. Using nitrous oxide instead of air as oxidant readily eliminates the net effect: the higher temperature decomposes the CaOH and eliminates the absorption band. Spectral interference due to scattering by atomization products can occur when the sample contains high concentrations of elements that form stable oxides (e.g. Ti or W). These metal oxide particles formed in the flame can cause light scattering. Fortunately, spectral interferences by matrix products are not widely encountered in FAAS and usually can be avoided by optimizing the flame composition or by adding interfering species to standards.

b) from molecular absorption or light scattering by gas combustion products. In this case, corrections are readily obtained from absorbance measurements made with a blank aspirated into the flame.

c) in instruments where the radiation source is not modulated spectral interferences may also arise by thermal emission of accompanying atomic lines and gas combustion products (see Figure 4.8) or by stray radiation.

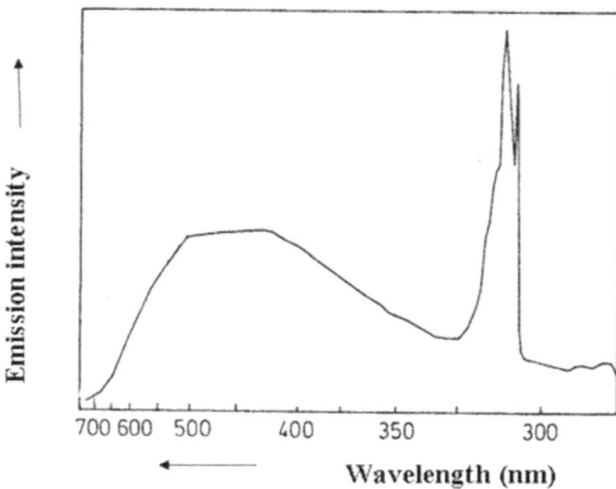

Figure 4.8. UV-Vis emission spectrum of an air/acetylene-flame showing the strong emission corresponding to the OH· bands in the 300 nm region and CH and "Swan bands" (C_2 molecules) between 300–600 nm.

4.4.2. Non-spectral interferences

For interferences other than spectral, the analyte signal itself may be directly affected. In this section, special attention will be paid to physical, chemical and ionization interferences.

a) **physical interferences** may be caused by differences in the physical properties of the sample and calibration standards, such as viscosity, surface tension, or density of the solution (for example, adding sulphuric acid makes the solution more viscous and decreases the absorbance, while the addition of methanol increases the absorbance signal by enhancing nebulization efficiency and thus increasing the amount of sample entering the plasma). They can be minimised by matching the density and viscosity of reference solutions and samples, for example by using the same solvent. Various authors have reported with a high degree of success the use of sampling pumps with fixed liquid flow rates to avoid these interferences

b) **chemical interferences** may be produced by the formation of a compound that prevents quantitative atomization of the analyte. There are a number of different approaches used in FAAS, which aid overcoming chemical interferences. Means commonly used are the selection of appropriate flame conditions, the matching of standard solutions to the sample, or the use of some special substances that reduces the interferences in specific chemical ways. In addition, in some circumstances, calibration by the standard addition method (see box in section 2.4 of Chapter 2) allows one to overcome such interferences.

As stated in Section 2.5 of Chapter 2, chemical interferences may be produced by formation of low volatility compounds, thus decreasing the rate at which the analyte is atomized. The most typical example is the effect of phosphate on Ca absorbance (phosphate reacts with calcium ions in the flame producing calcium pyrophosphate, which gives rise to lower absorbance signals than expected). These interferences can be reduced by adding matrix modifiers, such as "releasing agents", i.e. other cations that preferentially react with the anions (e.g. Sr or La will preferentially combine with phosphate and prevent its reaction with Ca), or "protecting agents" which form volatile species with the analyte and compete with interfering species (e.g., EDTA (ethylenediamine tetraacetic acid) complexes Ca and prevents interference from phosphate and sulphate). They may be eliminated also by using a higher temperature flame.

A serious situation occurs when the analyte reacts with flame gases forming thermodynamically stable oxides and hydroxides in the flame (e.g., aluminum, titanium and molybdenum). Several of these elements do not exhibit appreciable absorption in the conventional air-acetylene flame. To overcome this problem the flame composition has to be adjusted and the region of the flame monitored: a more useful flame for these elements is the nitrous oxide-acetylene flame in a reduction (fuel-rich) condition. In this case, the lack of oxygen-containing species combined with the high temperature of the flame, decompose and/or prevent the formation of refractory oxides.

c) **ionization interferences.** An appreciable fraction of alkali and alkaline earth elements and several other elements in hot flames may be ionized in the flame, giving rise to a decreased amount of atoms available for absorbance (a decrease of sensitivity and linearity is usually observed).

$$M^0 \rightleftarrows M^+ + e^-$$

Ionization can usually by detected by noting that the calibration line has a curvature upwards at higher concentrations, because a larger fraction of the atoms are ionized at lower concentrations. To overcome this problem, an "ionization suppressor" containing an easily ionizable element (B) can be added, e.g., potassium or caesium salts, provide a large amount of electrons to the flame and suppress ionization of the analyte. Thus, if the medium contains not only M but a significant amount of B species as well, and if B ionizes according to the equation:

$$B^0 \rightleftharpoons B^+ + e^-$$

then the degree of ionization of M is decreased by the mass-action effect of the electrons formed from B.

4.4.3. Calibration in flame atomic absorption spectrometry

Quantitative FAAS methods are based on calibration curves, which in principle are linear. However, departures from linearity can occur and analysis should never be based on the measurement of a single standard with the assumption that Beer's law is being followed over all concentration ranges.

The quickest and most practical method of calibrating an AA spectrometer is to prepare a series of calibration standard solutions (not by successive dilutions though). These solutions should match as far as possible the chemical and physical properties of the samples to be analyzed. The absorbance values of these standards are used to plot a calibration graph. In some cases these graphs may be curved and most commercial instruments allow the choice from a number of commonly used algorithms.

A widely practiced calibration method to overcome matrix interferences is the method of standard additions, which has been briefly explained in Chapter 2 (box in section 2.4). In any case, in spite of its broad applicability, here it should be pointed out that the standard additions method has some limitations, some of which are the following:

(i) the calibration graph must be substantially linear since accurate regression cannot be obtained from non-linear calibration points,

(ii) the fact that the measured part of the graph is linear does not always mean that linear extrapolation will produce correct results,

(iii) it is essential to obtain an accurate baseline from a suitable reagent blank,

(iv) it may require larger quantities of sample than other methods,

(v) the method can be very labor-intensive because every sample type requires its own set of calibration solutions,

(vi) operators must be carefully trained to prepare appropriate standards (this includes recognizing if the calibration is non-linear and preparing standards so the respective signals are in the linear region).

4.4.4. Analytical figures of merit

Table 4.2 lists some representative examples of detection limits (DLs) that can be achieved by FAAS (note that the DLs listed correspond to "instrument detection limits" or "IDLs" because a clean matrix was used. The "method detection limits" consider real-life matrices and usually are worse that IDLs). As can be seen in Table 4.2, the achieved DLs in FAAS depend greatly on the element analysed, and differences of three orders of magnitude can be expected depending on the analyte.

Under usual optimised conditions, the **precision** achievable by FAAS is of the order of 1 to 2%, but using special precautions this figure can be lowered to a few tenths of one percent.

Table 4.2

Examples of detection limits (DLs) obtained by FAAS ($\mu g.L^{-1}$). Adapted from "Guide to Inorganic Analysis". With permission of Perkin Elmer.

Element	DL	Element	DL	Element	DL
Al	45	K	3	Pb	15
Ca	1.5	La	3000	Ta	1500
Co	9	Mn	1.5	V	60
Cs	15	Mo	45	W	1500
Cu	1.5	Nb	1500	Zr	450

**All DLs were determined using elemental standards in dilute aqueous solution and instrumental parameters optimized for the individual element. Data shown were determined on an "AAnalyst™ 800".*

4.4.5. Use of organic solvents

When organic solvents are aspirated into a flame, an oxidizing (fuel-lean) flame must be used, because the solvent must be burned. The overall atomization efficiency for producing an atomic vapor in the flame is increased by the use of organic solvents in the sample solution. Such an increase may be due to a variety of causes, including increased rate of aspiration, finer droplets and more efficient evaporation, and/or combustion of the solvent. Thus, increased sensitivity is generally obtained. However, for safety reasons, except in some particular cases, the use of organic solvents in FAAS is falling out of practice.

Obviously much of our environment consists of water, therefore, the bulk of FAAS methods today deal with water as solvent. However, the use of organic solvents for AAS is necessary for certain applications, such as for solvent extraction of metal chelates and for the direct analysis of products like petroleum derivatives (a gasoline sample is too flammable to be introduced without dilution with a suitable miscible liquid into a flame instrument) or edible oils (an oil sample is too viscous to be aspirated without dilution). In such cases, particular care has to be taken with certain of an instrument's components because some materials used in the instruments employed for the analysis of aqueous solutions are often attacked by organic solvents.

There are a few solvents that should not be used in combination with FAAS like: halogenated hydrocarbons which readily decompose in the flame to produce phosgene (an extremely hazardous compound), very low boiling point hydrocarbons, ethers and acetone, tetramethylfuran and dimethylsulphoxide. Some of those solvents are so flammable that they can support a spectrometer flame without acetylene.

4.5. APPLICATIONS AND CASE STUDIES.

FAAS is today a well-established technique for trace metals analyses in almost any human activity area and for a huge variety of different samples. Therefore, it is not easy to select a few practical and illustrative examples of FAAS applications. Many FAAS standard procedures established by international standard committees are now routinely used. Besides, most companies manufacturing commercial instruments provide tutorial books and application notes. The examples finally selected below were chosen from three very different areas and we will just try to give some guidelines for laboratory work.

4.5.1. Determination of calcium in milk.

Ca is one of the elements most frequently analyzed by FAAS. The determination of calcium in milk requires the removal of milk proteins including casein, for example by precipitation using trichloroacetic acid. The samples are then filtered and the filtrate analyzed by FAAS (in alternative procedures the organic part of milk is removed by digestion with nitric acid and hydrogen peroxide in a microwave oven).

The slight Ca ionization that occurs in the air-acetylene flame can be controlled by the addition of an alkali salt (0.1% or more potassium as chloride) to both samples and standards. Calcium sensitivity is reduced in the presence of elements that give rise to stable oxy salts, including Al, Si, Ti, P, V and Zr. This effect is reduced by the addition of 0.1-1.0% La or Sr or 1% EDTA (as di-sodium salt).

The absorption of calcium is dependent on the fuel/air ratio and the height of the light beam above the burner. Although maximum sensitivity is obtained with a reducing (fuel-rich) flame, an oxidizing (fuel-lean) flame is frequently recommended for optimum precision. Calcium determination appears to be free of most chemical interferences in a nitrous oxide-acetylene flame; in this case, only ionization interferences need to be controlled, and this can be easily done by the addition of an alkali salt as mentioned above.

4.5.2. Determination of molybdenum in fertilizers.

Sandy soils and those that are inherently infertile in their natural state, e.g., soils low in phosphorus, are also typically low in molybdenum. However, Mo is an essential micronutrient in plants (Mo is important in nitrogen metabolism and the synthesis of proteins). Therefore, fertilizers containing Mo, which can be applied to the soil or the foliage, are commercially available.

Elements forming refractory oxides in flames (such as Al, Ti, W, Cr and Mo) require a high temperature flame to produce adequate atomic populations. Thus a nitrous oxide-acetylene flame has to be used, but with attendant experimental difficulties (this flame can be dangerous if not operated properly as its burning velocity is relatively high). Unlike most of the refractory elements (with the exception of Cr, which behaves like Mo), molybdenum can be determined also in a lower temperature air-acetylene flame, in addition to the more often used high temperature nitrous oxide-acetylene flame. In the former case, fuel-rich conditions are required to compensate for the lower temperature in order to obtain an adequate population of Mo atoms. In any case, with the nitrous oxide-acetylene flame, a better sensitivity (about an order of magnitude improvement) can be achieved due to the higher flame temperature and more favourable flame environment and, at the same time, selection of conditions with this flame is not so critical as for the air-acetylene flame. Besides its higher sensitivity, the nitrous oxide/acetylene flame yields much less interferences, so that it is used almost exclusively nowadays.

Interferences in the determination of molybdenum in fuel gas rich air/acetylene flames have been scarcely investigated and matrix matching is frequently used. Although a number of interferences may occur in a nitrous oxide/acetylene flame (e.g., calcium, strontium, iron and sulphate), such interferences can be controlled by the addition of 0.5% aluminium chloride (aluminium seems to inhibit the lateral diffusion of molybdenum atomic vapour, leading to its increased atom concentration in the middle of the flame).

4.5.3. Determination of lead in gasoline

The determination of metals by FAAS in carbon-rich samples such as edible oils, motor oils, lubricating oils, and petroleum products are based on char-ashing or dilution of the sample with an organic solvent. This char-ashing technique affords accurate results for some elements, while allowing the determination of trace metals at a much lower level than direct sample aspiration. The main disadvantage of this method consists in its tedious and lengthy procedures, since the sample must be first completely carbonized on a hot plate before it is ashed in a muffle furnace. Moreover, it cannot be applied to volatile analytes, which would be lost during sample preparation.

A typical FAAS determination is lead in gasoline. In this case, the method of sample preparation involves the reaction of the alkyl lead components of gasoline with iodine (the reaction with iodine reduces the variation in response for different alkyl lead compounds). Then the stabilization of the alkyl lead iodide complexes with tricaprylmethylammonium chloride (Aliquat 336) and dilution with methyl isobutyl ketone (the order of addition of the reagents must be followed explicitly: Aliquat 336 must not be added before the addition of iodine because it retards the addition of iodine and the formation of the alkyl lead iodide-Aliquat 336 complex, bringing about erroneous results).

SAFETY PRECAUTIONS.

1. Use of FAAS with gasoline samples requires complete adherence to all relevant safety practices when flammable materials are analysed in a flame.
2. Antiknock lead compounds are particularly poisonous and must not be inhaled or ingested or come into contact with the skin. Also, they should never be exposed to elevated temperatures (above 50°C) or to acids and oxidizing agents.
3. Solvents used in this methodology present a hazard risk to users. Therefore, operators should consult the relevant Material Data Sheet before their use.

Chapter Five

HYDRIDE GENERATION AND COLD VAPOR ATOMIC ABSORPTION SPECTROMETRY

5.1. INTRODUCTION

The generation of volatile derivatives of an analyte, as a preliminary step for a more efficient analyte introduction into an atomizer, is now a well-established sub-discipline in AAS. As mentioned in Chapter 4, typical detection limits obtained with conventional flame AAS (FAAS) are of the order of tenths of mg.L^{-1} (ppm) or even, in some analyses, at the ppm level. Unfortunately, these detection limits are very high for many practical applications, such as those needed for the determination of many toxic metals in biological samples or in the environment. Three main alternatives can be considered to improve the sensitivity of AAS determinations: (i) to employ a different atomizer than a flame, in particular, electrothermal atomization in a graphite furnace (Chapter 6 is addressed to this topic), (ii) to use an appropriate analyte pre-concentration system (e.g., liquid-liquid or solid-liquid), which will be dealt with in Chapter 7, or (iii) to resort to analyte vapor generation (VG) techniques.

VG coupled to AAS comprises three main stages: (i) generation of the volatile analyte derivative from the liquid (or solid) sample and its transfer to the gaseous phase, (ii) its collection (optional) and transfer to the atomizer by a flow of a purge gas and, (iii) finally, its decomposition in the spectrometer light path to provide the gaseous metal atoms. VG allows for a high transport efficiency of analyte into the atomic absorption cell and this is a most significant advantage in comparison with conventional FAAS: in conventional FAAS the nebulization process limits the sample introduction rate; further, only a small fraction of the sample nebulized (e.g., 5-10%) ever reaches the flame, with the remainder being directed to drain. Furthermore, the sample introduced into the flame remains in the light path for a very short period ("residence time") as it is propelled upwards through the flame.

However, in VG-AAS all the volatilized analyte is transported to the measurement absorbance cell, which usually has a longer long light-path and also provides for a longer residence time in the atomizer, thus increasing analyte signals.

Various chemical systems for VG have been reported in analytical atomic spectrometry for up to 30 elements, including generation of volatile fluorides, chlorides, carbonyls, hydrides, atoms, but only a few of these systems offer competitive performance characteristics [**Tsalev** (**1999**)]. Most attractive features of these chemical vapor generation systems are: (i) reactions in aqueous media, (ii) fast kinetics of VG and easy stripping-off of vapors from solution, (iii) reliable interference control, (iv) high and reproducible chemical yields of volatile products, which are (v) readily atomized in a quartz tube or graphite furnace atomizer rather than in a flame.

In this context, hydride generation (HG) of metal(loid)s such as As, Bi, Ge, Pb, Sb, Se, Sn or Te and cold vapor (CV) formation of Hg and, more recently, Cd, are by far the most widespread applications of VG for AAS detection.

Numerous volatile and semivolatile metal compounds are present in our environment as a consequence of both anthropogenic and natural processes [**Sturgeon and Mester** (**2002**)]. For example, metal hydrides can be formed in many places where reducing conditions prevail and several reports have highlighted detection of metal hydrides arising from landfill and waste-water treatment sites. Formation of metal hydrides in nature is primarily considered to be the result of bacterial activity. Other major sources for the volatilisation of trace elements include biotic and abiotic alkylation (usually methylation) processes. It has been also reported that the presence of volatile molybdenum and tungsten carbonyl species exists in municipal waste deposit sites. Also, a less known organometallic pollutant, the gasoline additive methylcyclopentadienylmanganese tricarbonyl, has been detected in various environmentally sensitive places. On the other hand, the formation and release of elemental vapors also occurs in the environment: ionic mercury can be reduced to atomic mercury and be widely dispersed.

5.2. VOLATILE HYDRIDES GENERATION BY TETRAHYDROBORATE (III) IN AQUEOUS MEDIA

Formation of volatile hydrides offers a practical and sensitive way for the determination of several important elements that suffer problems when determined by straightforward nebulization AAS methods. In particular, the hydrides of Groups 14-16 (IVB-VIB) in the periodic table are of particular importance to the practical AAS determination of As, Bi, Ge, Pb, Sb, Se, Sn, Te and, to some extent, In and Tl. In all these cases, FAAS limits of detection are readily improved by several orders of magnitude by generating volatile hydrides.

The early chemical systems for volatile HG used a metal-acid reduction system, such as Zn-HCl. However, this reaction is limited to a few analytes, is slow, difficult to automate, suffers from high blank signals due to zinc metal impurities, and it is not usually efficient (as a result of incomplete reaction and/or entrapment of the hydride in the precipitated zinc sludge).

Nowadays, the almost universally system used for hydride generation is based on sodium or potassium tetrahydroborate (THB) in an acidic medium. This approach allows for a fast derivatization process directly from aqueous media.

5.2.1. Mechanism of hydride formation

The reaction mechanisms related to the interactions of THB in aqueous solution with ionic or molecular species of analyte elements, which are the precursors to the volatile derivatives of interest, are not well known yet. To describe the mechanism of HG it is usually postulated that the active species in the reduction process is atomic hydrogen or "nascent hydrogen", which is thought to be formed during the acid hydrolysis of THB:

$$BH_4^- + H^+ + 3H_2O \longrightarrow H_3BO_3 + 8H$$

with the atomic hydrogen being responsible for the formation of the hydride:

$$M^{m+} + (m+n)H \longrightarrow MH_n + mH^+$$

where m corresponds to the oxidation state of the analyte and n is the coordination number of the hydride. The excess unreacted atomic hydrogen forms molecular hydrogen, through reactions of the type:

$$H + H \longrightarrow H_2$$

$$H + H_2O \longrightarrow H_2 + OH$$

However, the formation of nascent hydrogen by decomposition of THB during acid hydrolysis is not yet understood. The attack of protons on the THB and on its intermediates (trihydroboron, dihydroboron and monohydroboron species) generates molecular hydrogen. However, none of the numerous kinetic and mechanistic studies reported in the fundamental chemistry literature on this topic documents or even hypothesizes the existence of nascent hydrogen intermediates [**D'Ulivo (2004)**].

The most important and sound information on the mechanism of hydride generation stems from studies that have been performed on a few elements using deuterated reagents. For example, in studies on arsine generation using deuterated reagents and mass spectrometry, it was found the reaction of both As(IIII) and As(V) with $NaBD_4$ in 3M HCl and H_2O produces AsD_3 as the main product. When the reaction was performed using $NaBH_4$ and arsenic species in DCl and D_2O, the main product was AsH_3. In other words, at elevated acidities, arsine is produced by direct transfer of hydrogen to the analyte from boron. This is in contrast to the nascent hydrogen mechanism, which should give rise to all the possible randomly deuterated species AsH_nD_{3-n} (n = 0, 1, 2, 3). This evidence supports a mechanism based on the direct transfer of hydrogen atoms from borohydride reagent to arsenic and is of particular relevance because it suggests that some active species containing boron-hydrogen bonds could still exist in solution, even at elevated acidities, where the rate of decomposition of THB is expected to be very fast. This observation, combined with the mechanism of hydrolysis of THB in acid solution results in the conclusion than in HG an important role is played by the BH_4^- intermediates of trihydroboron, dihydroboron and monohydroboron species, which appear to be more resistant to hydrolysis than BH_4^- itself, under strongly acidic conditions [**D'Ulivo (2004)**].

To conclude this section it is important to point out that the actual mechanism of the HG reduction reaction still remains to be better clarified. However, from an analytical point of view this is not of crucial importance since, in practice, there is always several order of magnitude excess of THB over analyte levels in the sample.

5.2.2. Basic instrumentation

The chemical VG procedure can be carried out either in batch or in flow mode [**Inczédy et al (1998)**]. In both cases, the gaseous species generated are introduced, with a stream of a purge inert gas, into the instrument's optical path for measurement by AAS:

(a) In the *batch generation mode,* specified volumes of the sample and reactant solutions are mixed at the beginning of the measuring process in a batch generator. A batch generator (see Figure 5.1) is a vessel made of glass or plastic serving both as the reactor as well as the separator of gases from the reaction mixture. Gas liberation in the generation vessel is accelerated by stirring, agitating or by passing an inert purge gas (e.g., nitrogen or argon) through the solution. Released hydride is thus transported by the purge gas flow to the atomizer together with the hydrogen formed from reducing agent decomposition. After completion of the hydride evolution, the reacted mixture has to be disposed to waste, the generator rinsed and a new batch or sample is then added. The analytical AAS signals obtained have a peak-shaped response *versus* time. The sensitivity can be further increased by using larger sample volumes: since all of the analyte contained in the sample is released for measurement, increasing the sample volume means that more atoms of analyte are available to be transported to the absorbance cell. However, it has to be noted that the relationship "sample volume:sensitivity" is not linear (at high volumes the sensitivity is not as high as, in principle, expected) and this has been attributed to factors such as higher solubility of the hydride when the sample volume is high and also to hydride losses due to interaction of the hydride within a larger reaction flask.

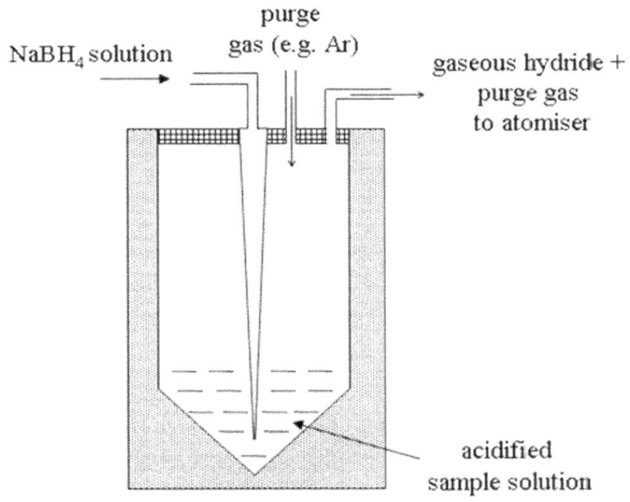

Figure 5.1 Batch chemical vapor generation system.

(b) In the *flow mode* the sample, either injected as a discrete volume or continuously flowing through a flow system channel, mixes with a flow of reductant. As can be seen in Figure 5.2, additional channels may be used to prepare the sample. The mixed stream is connected on-line to a gas-liquid separator: the generated gaseous species are then transported to the atomizer (also connected on-line) while the liquid goes to a waste. In the first case (flow injection mode) a peak-shaped signal *versus* time is achieved, while in the second (continuous mode) a steady-state signal is obtained. Advantages brought by the use of the flow injection mode are lower sample memory effects (the carrier stream cleans the flow-system and the atomizer between sample injections) and a higher sample throughput.

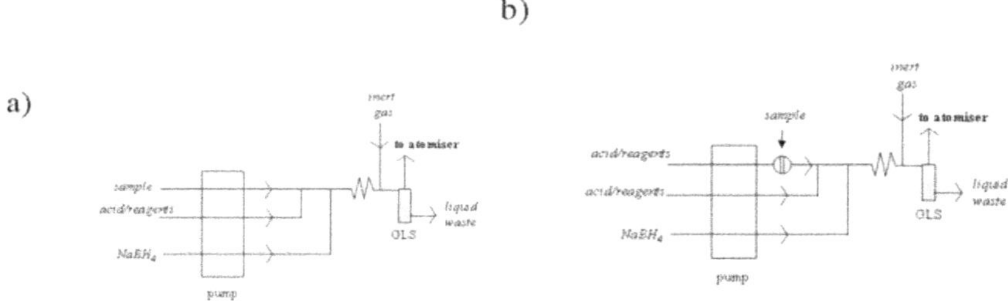

Figure 5.2. Flow configurations for hydride generation. GLS: gas-liquid separator. (a) Continuous mode. If the sample was previously conditioned off-line with acid and reagents, the corresponding channel in the flow system is not needed, and only two channels are then necessary: one for the conditioned sample and the other for $NaBH_4$. (b) Flow injection mode. Usually the sample volume injected is low (e.g. 100 μL) or has been conditioned off-line (acid and reagents); in such cases, the intermediate merging channel in the diagram containing acid/reagents is not needed.

Flow systems for the determination of volatile derivatives of the analyte(s) are very commonly used nowadays and they are described in detail in section 7.6 (Chapter 7)

Currently, the most often employed atomizers in HG-AAS are heated quartz tubes and graphite furnaces located in the optical path of the atomic absorption instrument. Figure 5.3 shows a schematic of a typical quartz cell, usually made of clear fused quartz tubing, with a "T" configuration (a central inlet tube for the hydride and the carrier gas, a long light-path, and two gas exits). The quartz absorption cell can be heated by a flame (the cell is mounted over the burner head of the spectrometer) or by an electric cell heater. In either case, the hydride gas is dissociated in the heated cell (e.g., 800-1000°C) giving rise to free atoms of the analyte.

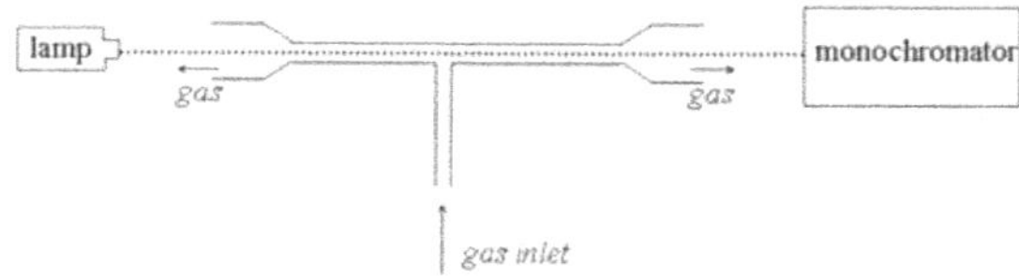

Figure 5.3. Quartz cell used as atomizer in HG-AAS.

Finally, it is important to note that nowadays most manufacturers of instruments for AAS commercialize accessories allowing for both batch and flow vapor generation.

5.2.3. Limits of detection

Table 5.1 shows a comparison of detection limits (DLs) obtained for several volatile hydride forming metals using different optical spectrometric techniques: conventional FAAS, HG coupled to AAS, ETAAS and ICP-OES. As can be seen, DLs achieved by HG-AAS are several orders of magnitude lower than those with conventional FAAS. Also they compare favorably with those obtained by ICP-OES and, in most cases, are of the same order of magnitude as DLs obtained by ETAAS.

ELEMENT	FAAS[1]	HYDRIDE-AAS[2]	ETAAS[3]	ICP-OES[4]
As	150	0.03	0.05	2
Bi	30	0.03	0.05	1
Sb	45	0.15	0.05	2
Se	100	0.03	0.05	4
Te	30	0.03	0.1	2

[1] All DLs obtained by AAS were determined using instrumental parameters optimized for the individual element, including the use of electrodeless discharge lamps where available. Data shown were determined on an "AAnalyst™ 800".
[2] Hydride detection limits shown were determined using an "MHS-15 Mercury/Hydride system".
[3] ETAAS detection limits were determined on an "AAnalyst 800" using 50 □L sample volumes and an integrated platform.
[4] All ICP-OES ("Optima 4300/5300)" DLs were obtained under simultaneous multielement conditions with the axial view of a dual-view plasma using a cyclonic spray chamber and a concentric nebulizer.

Table 5.1 Atomic spectroscopy detection limits (DLs), in $\mu g.L^{-1}$, for five selected analytes. Adapted from "Guide to Inorganic Analysis". All DLs were determined using elemental standards in dilute aqueous solution and are based on a 98% confidence level (3 standard deviations). With permission of Perkin Elmer.

To account for the important improvements obtained in comparison with conventional FAAS, the factors pointed out in the introduction of this chapter have to be taken into account: in HG the inefficient sample nebulization step is not needed, the residence time of the analyte in the optical path is larger and the optical-path longer. Moreover, an additional factor accounting for the sensitivity increase in comparison with conventional FAAS, is that the background is much lower since most of the analytes have the resonant lines in the region of 200 nm where the flame produces a considerable background, while when the HG procedure is carried out the analyte volatile hydride is atomized in a flame-free environment (argon or nitrogen are the gases in the optical cell).

5.2.4. Selectivity: sources of interferences

Interferences in HG may occur either in the liquid phase, during hydride formation and its transfer from solution (e.g., decreased efficiency of hydride formation or changes in the rate of hydride release from the liquid phase), or they can affect the analyte in the gaseous phase, either during transport of hydride from the sample solution to the atomizer or in the atomizer itself.

Interferences in the liquid phase can be due to different reduction efficiencies of different analyte species in the sample or to the sample matrix. In the first case, adequate precautions should be taken in the sample preparation stage, like appropriate pre-reduction or sample digestion. For example, pre-reduction is almost indispensable for Se and Te, the higher states of which do not produce hydrides with THB. Also, in As and Sb determinations, while both trivalent and pentavalent forms of analytes yield hydrides, nevertheless pre-reduction is highly desirable since arsenite and antimonite are determined with better sensitivity (2-8 fold). On the other hand, organo-element species of an analyte usually need to be decomposed prior to the HG step, generally by wet chemical attack with strong oxidants at room or elevated temperature (microwave or UV-assisted).

In respect of the sample matrix, care should be taken about the possibility of solid particles capturing analyte ions (e.g., carbon particles from an inefficient mineralization of biological samples), and also with transition ions and noble metals that form species by reaction with THB able to capture the hydride. In most cases, the matrix species responsible for such interferences are usually present in great excess compared with the analyte. Thus, the extent of the interference should not depend on the analyte concentration but only on the interferent concentration and, therefore, the method of standard additions may be used, in principle, to overcome these interferences. In other cases (in particular when the analyte signal is seriously suppressed) interference has to be masked or physically removed prior to the reduction stage.

Other volatile species, most often hydrides, but also other compounds, or liquid spray produced in the hydride generator, may produce transport interferences in the gas-phase; however, there are not many cases of these interferences reported in the literature. Another source of gas-phase interferences are due to interaction and decomposition of the hydride on the surface of the apparatus connecting tubes or even on the atomizer surface itself very often giving rise to memory effects.

5.3. ELECTROCHEMICAL GENERATION OF VOLATILE HYDRIDES

Although chemical reduction with sodium tetrahydroborate is by far the most common method for HG–AAS determinations, the use of THB for such purposes is not ideal, as some drawbacks of this reagent include [**Sturgeon and Mester (2002)**]: (i) THB is prone to undesirable contamination (worsening the limit of detection of analytes determination) and is relatively expensive, (ii) its aqueous solutions are unstable and, so, should be prepared immediately prior to use (even if some improved stability can be conferred through addition of 0.1-2% NaOH, membrane filtration or refrigeration), (iii) the corresponding derivatization reaction may be interfered with by concomitants in the sample solution, (iv) evolving excessive amounts of hydrogen may alter the performance of some detection systems.

Because of these problems, electrochemical hydride generation (EcHG) has been investigated for analytical purposes in recent years [**Sima *et al* (2004), Bolea *et al* (2006)**]. It is important to note, however, that EcHG is not yet established for routine analyses by AAS. In a typical EcHG experiment, continuous flows of catholyte and anolyte (electrolyte adjacent to a cathode or anode, respectively) are pumped through the cathode and anode channels, respectively of an electrolytic flow-through cell. The sample is typically injected into the catholyte stream used as carrier (e.g., diluted hydrochloric or sulphuric acid). In this way, the hydride is generated in the cathodic space of an electrolytic cell with concurrent oxidation of water in the anodic compartment, while the gaseous reaction products are then separated from the waste aqueous solution in a subsequent gas-liquid separator using a purge gas flow. The main advantage of the EcHG approach is that it does not require any other reagents than diluted acid, this brings also as an additional advantage that the risk of contamination is reduced since these acids can be obtained in high purity. Additionally, the experimental conditions are not as critical and the influence of the oxidation states of the analyte on the hydride yield can be reduced. Also, it has been reported that interfering levels in EcHG are lower in general terms than those observed in THB-based generation systems. However, interferences still exist. Using EcHG, the major interference source does not occur in solution, as in chemical vapour generation systems, but on the surface of the cathode. As a consequence, masking of interfering ions is usually ineffective in preventing interferences in EcHG systems. Most of the authors proposing remedial actions for transition metal interference elimination using EcHG have resorted to procedures of physical separation of the matrix and/or pre-treatments of the cathode surface [**Bolea *et al* (2006)**].

Several cathode materials have been investigated for EcHG. The most successful ones seem to be carbon-based (glassy or pyrolytic graphite) and Pb [**Arbab-Zabar et al (2006)]**. An often-reported construction of electrolytic cell consists of two-half cells, an anodic and a cathodic one, usually separated by an ion exchange membrane or porous glass frit. Figure 5.4 shows a design of a typical electrolytic cell constructed by assembling four blocks, made of Perspex®, held together with screws. A coil of Pt wire and a lead-tin alloy wire, both having a surface area of 1.0 cm^2, were used as anode and cathode, respectively. An argon carrier gas flow is directly introduced into the cathode chamber for the separation of the generated volatile species. Thin layer miniaturized flow-through cells with sheet-shaped electrodes have been also proposed [**Sima et al (2004)**] allowing low flow rates of catholyte.

EcHG-AAS methods have been successfully applied to trace level determinations, for example, of arsenic, antimony, selenium, tin, bismuth, germanium, cadmium.

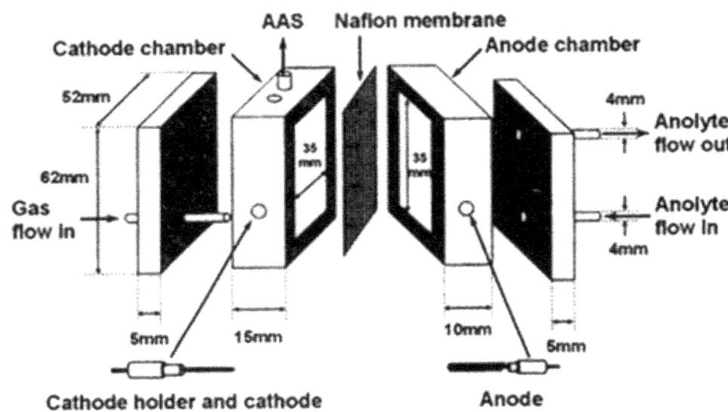

Figure 5.4. Diagram of a electrolytic hydride generation cell for electrochemical generation of volatile hydrides (from M.H. Arbab-Zavar, M. Chamsaz, A. Youssefi, M. Aliakbari, Anal. Chim. Acta, 576 (2006) 215. Reproduced with permission of Elsevier).

5.4. COLD VAPOR GENERATION

Since free atomic vapor for most metals cannot exist at room temperature, heat must be applied to the sample to break the bonds that bind atoms within molecules and to obtain the atoms in a vaporized form (necessary for AAS measurement). A notable exception to this general requirement that occurs in nature is mercury: vaporized free mercury atoms can exist at room temperature and, therefore, mercury can be measured by AAS without the need for a heated absorbance cell. Cold vapor (CV) generation for determination of mercury was first described in the 1960s, after which it became the leading method for mercury determination. Also, more recently, in 1995, two independent research groups [**Sanz Medel *et al* (1995), Guo *et al* (1995)**] reported on the AAS determination of cadmium following an appropriate reaction of the metal aquo-ion in acidic medium with THB to form CV generation of cadmium for the final AAS measurement.

5.4.1. Mercury

Mercury CV generation is based on the reduction of mercury ions with a reducing agent to generate elemental mercury. Since the vapor pressure of elemental mercury is quite high, Hg^0 can be readily separated from the aqueous matrix with an inert purge gas (e.g., Ar) and then transported and introduced into an absorbance cell. As pointed out before, the analyte reaches the absorbance cell already in atomic form, and so there is no need for an additional atomization step. However, it is usually advisable to heat the cell, in order to avoid background absorption from water vapor transported by the carrier gas, which at room temperature may condense as small droplets in the absorption cell.

Either tin chloride or sodium tetrahydroborate are used as a mercury reductant. An advantage brought by the use of tin chloride is that hydrogen gas is not evolved. This fact is not so important in AAS, but may be of particular concern when plasma-based detectors are used (because hydrogen produces plasma instabilities and so poor precision, e.g., in ICP-OES measurements). However, it should be pointed out that reduction of Hg^{2+} by tin chloride is slower than that using THB. Moreover, this reagent is unable, by itself, of reducing organo-mercury species; therefore, in this case, the CV determination of total mercury is carried out using a wet digestion process with oxidizing reagents, in order to decompose any organic mercury to Hg^{2+}. Conversely, sodium tetrahydroborate is capable of reducing both inorganic and organo-mercury species. However, the resulting gaseous products are different, with inorganic mercury being reduced to elemental mercury, while the existing organo-mercury species are transformed to the corresponding volatile hydride species.

5.4.2. Cadmium

Recently, Cd has been shown to be able to form monoatomic vapor at room temperature that is stable enough to be measured by AAS. In fact, CV-AAS has been used advantageously for the determination of Cd in a variety of matrices via reduction of the cadmium ions in solution by $NaBH_4$. It has been reported that the CV reaction efficiency is dramatically enhanced in the presence of organic reagents such as thiourea and/or "organised media" (see text box below), to the extent that the reported overall efficiencies for generation of atomic cadmium reached 75%.

The term "organised media" has been applied in Chemistry to describe microscopically organised chemical assemblies, e.g., of a surfactant (amphiphile), which spontaneously form in the bulk of a solution. Organised media or "organised assemblies", such as micelles and vesicles, have been shown to exhibit valuable properties in analytical chemistry: they can solubilize, concentrate and compartmentalize ions and neutral molecules, they can modify equilibria, e.g., acid-base and redox constants, reaction rates, chemical pathways and also influence stereochemistry. In addition, they may drastically change the spectral characteristics of coloured/luminescent compounds **[Fernández de la Campa *et al* (1995), Burguera *et al* (2004)].**

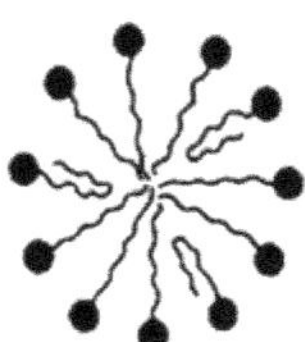

Micelle: self-aggregation of individual surfactant molecules above a critical concentration.

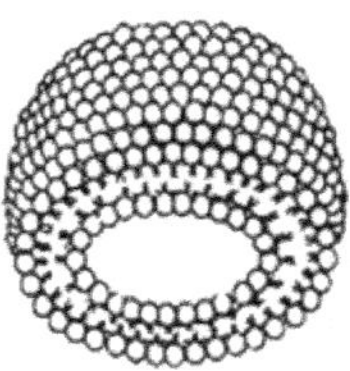

Vesicle: closed bilayer structure formed by double chain surfactants after adequate sonication.

Although the reaction mechanism is not fully understood yet, it seems plausible that the reaction intermediate may be CdH_2, transportable over significant distances at room temperature where, under the influence of the UV excitation from the atomic absorption instrument's lamp, it ultimately would decompose to yield the measurable Cd^0 [**Sanz Medel et al (1995)**].

5.5. TRAPPING/PRECONCENTRATION SAMPLE INTRODUCTION

Volatilized metal-derivatives can be directly transported into the light path ("direct transfer systems") for analysis. However, the detection sensitivity can be improved by first trapping (and preconcentrating) the volatilized analyte derivative "in or on" an appropriate medium ("collection device"). Then, after VG evolution has finished, the analyte should be rapidly released from the collection device. A variety of collection methods were most frequently used in the early years of the application of CV-AAS when relatively slow approaches for volatile-species formation were employed.

For hydrides, at present, only cryogenic trapping (located prior to the atomization cell) and, most commonly, *in situ* trapping in a graphite furnace (here the graphite furnace acts also as an atomization cell) are being used. In both cases, hydrogen evolved in the reaction passes freely and, so, is not collected:

a) **Cryogenic trapping.** One of the main advantages of cryogenic trapping compared to other preconcentration methods is that unstable/reactive species do not come into contact with any liquid or solid sorbent material, significantly reducing the possibility of their alteration/decomposition. Typical trapping temperatures are in the -150 to -196 ºC range, obtained using liquid nitrogen cooling. After the collection stage is complete, analytes are released following fast heat-up.

However, cryotrapping technology remains rather unpopular: most cold traps are home-made and are operated manually, requiring skill and experience of the operator. In any case, it should be highlighted here that the approach is particularly used on-line coupled to gas chromatography (GC) in the metal-speciation field, for preconcentration of volatile metal(loid) species obtained by hydride generation/ethylation derivatization, to be separated afterwards by GC.

b) ***In situ* trapping.** Major impediments to enhancing performance of conventional VG-based analytical techniques are that the species are often diluted by the co-evolved hydrogen as well as any gas used to ensure phase separation and transport. Further, some hydrides are not efficiently atomized in heated quartz tube sources (e.g., GeH_4). Both inconveniences can be solved by resorting to *in situ* trapping techniques, which couple HG with a graphite furnace and ETAAS determination, thus permitting significant enhancement in detection capabilities over conventional batch and flow generation procedures. In this approach the graphite furnace (located in the optical-path) is used to decompose the volatile hydride and trap the analyte species on the tube surface, or onto a surface pretreated (e.g., with palladium or iridium), being later released by a controlled furnace heating-program.

For the case of mercury, a collection device based on amalgamation is commonly used to measure samples with very low mercury concentrations (see Figure 5.5): mercury vapor liberated from the sample aliquot in the reduction step is trapped on a gold (mercury readily forms an amalgam with gold) or gold alloy gauze. After a suitable collection time, the gold surface is then electrically heated and the absorbed mercury released, the vapor being directed to the absorbance cell (in some designs the mercury amalgamation trap is fitted into the quartz tube). The only theoretical limit to this approach would be that imposed by background or contamination levels of mercury in the reagents or system hardware. Nowadays, several companies manufacture compact, automated atomic absorption instruments dedicated to the determination of mercury, with amalgamation collection system as an option.

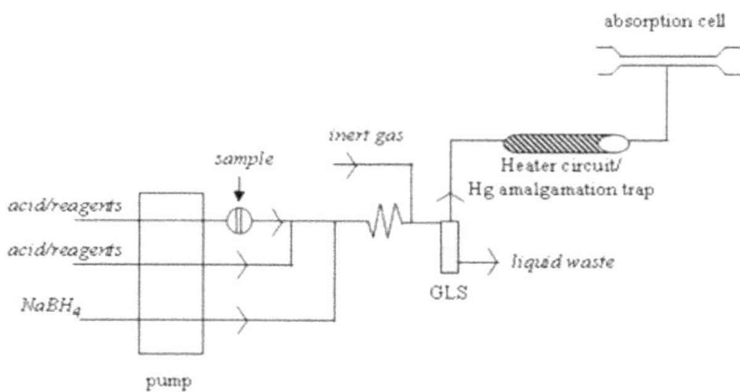

Figure 5.5. Schematic diagram showing the key components of a flow injection system with mercury amalgamation trap assembly for mercury determination by CV.

5.6. APPLICATIONS AND CASE STUDIES

Hydride and cold vapor generation accessories are today commercially available and relatively inexpensive and offer low detection limits, the possibility of automation (in flow modes) and good sampling frequencies (e.g. 40-60 h^{-1}). In addition, as it can be seen in Table 5.2, HG offers certain (limited) potential for analytical metal speciation.

On the other hand, HG and CV methods are not free from intrinsic limitations and drawbacks. A major limitation is that, in practice, they are restricted to a few analytes. Also, experimental parameters need to be carefully controlled and care should be taken with chemical interferences. Moreover, specific chemical pre-treatment is often needed so as to reliably transform the analyte element into a definite, reactive chemical form and oxidation state [**Tsalev (2000)**]. Thus, for example, selenium and tellurium analytes have to be pre-reduced to their hydride-forming oxidation states, Se(IV) and Te(IV); lead should be pre-oxidized to its metastable Pb(IV) form; arsenic and antimony usually exhibit much better sensitivity as their inorganic As(III) and Sb(III) oxidation states and a pre-reduction step is often necessary.

SPECIES DETERMINED	DIFFERENTIATION FROM:	PRINCIPLE
i-Se(IV)	i-Se(VI)	Reaction kinetics
i-Te(IV)	i-Te(VI)	Reaction kinetics
i-As(III)	i-As(V), MMA, DMA	pH; $NaBH_4$ concentration
i-As(III) + DMA	i-As(V), MMA	pH
i-As(III) + i-As(V) + MMA + DMA	AsBet, Me_4As^+, arsenosugars	Reaction kinetics

i: inorganic.
MMA: monomethylarsonic acid.
DMA: dimethylarsinic acid.

Table 5.2 Examples of HG-AAS speciation potentialities (adapted from ***Tsalev 2000****). With permission of Elsevier.*

Today, HG-AAS for As and CV-AAS for Hg are both well established, highly effective techniques for these two difficult analytes. Thus, in order to illustrate these two typical AAS determinations where improved sensitivity is achieved by resorting to volatile species generation, the determination of arsenic in water and of mercury in hair will be described in more detail. Of course, the high toxicity of both metals even at concentrations down to $\mu g.L^{-1}$ (ppb) justifies the selection of such illustrative applications.

5.6.1. Determination of arsenic in water

Human exposure to arsenic can cause a variety of adverse health effects. Living organisms in general may be exposed to toxic arsenic species primarily from food and water. The contamination of groundwater with arsenic has been reported in several countries around the world. In particular, fifty districts of Bangladesh and nine in West Bengal (India) have arsenic levels in drinking water well above the guideline values (10 $\mu g.L^{-1}$) provided by the World Health Organization (WHO). According to the WHO in the year 2000: "the contamination of groundwater by arsenic in Bangladesh is the largest poisoning of a population in history". Hyperkeratosis on the palms and feet is the common symptom of arsenic poisoning. Long-term exposure to low concentrations of arsenic has been reported to cause cancer of the bladder, skin and other internal organs.

In groundwater, arsenic is predominantly present as arsenite and arsenate with a minor amount of monomethylarsonic acid and dimethylarsinic acid (see corresponding structures in Figure 5.6). Under appropriate experimental conditions the four species can be reduced to As volatile hydride with THB, although transition metals might interfere with this sensitive determination of arsenic. The predominant mechanism is probably due to the reaction of the interfering transition metal ion with $NaBH_4$ reductant, forming a precipitate, which is able to encapsulate and/or catalytically decompose the hydrides. Generally L-cysteine has proved to be very useful for preventing iron interferences, commonly present at high concentration in many types of samples. Moreover, it was found that a cysteine-THB intermediate formed showed higher reducing power than THB alone.

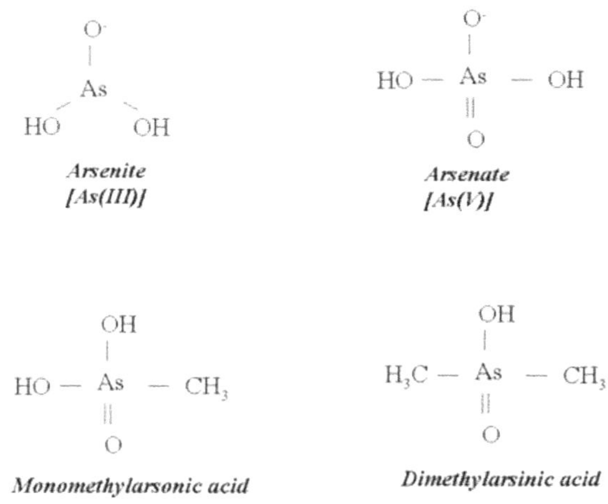

Figure 5.6. Arsenic species in water.

HG can be also used for differential determination of As(III) and As(V), just by taking advantage of the fact that As(III) reacts with THB at a higher pH than As(V). It should be noted, however, that the analytical potential of this approach for chemical speciation in general, and for As speciation in particular, is rather limited. Thus, when speciation of all possible arsenic compounds present is required, the combination of a more powerful separation method (e.g., chromatography) prior to HG-AAS is almost mandatory (see section 7.10.3 in Chapter 7). Speciation information on arsenic in natural waters is very important today because the bioavailability and the physiological and toxicological effects of arsenic will depend on its chemical form (for instance, arsenite is 20-60 times more toxic than arsenate and several hundred times more toxic than methylated arsenic compounds).

The lack of long term stability of arsenic species in water samples is a crucial aspect in As and other speciation studies. Procedures used for sample storage, handling and analysis may result in final oxidation of arsenite to arsenate. In groundwater, As(V) has commonly been reported as the predominant water-soluble As species. However, recent studies have shown that As(III) also may be present in high concentrations in certain waters. Therefore, particular attention should be paid to potential changes of sample conditions when transporting it from the field to the laboratory and during storage, because alterations of chemical species in the original sample might happen changing the original species present in the water. If no reliable sampling and storage conditions can be found, on-site analysis with portable instruments may be necessary.

5.6.2. Determination of mercury and methylmercury in hair

The first well-documented cases of severe poisoning by mercury compounds are from Minamata Bay (Japan) in 1956 when sea fish were severely contaminated with methylmercury (coming from an industrial spill). Later, other occurrences of poisoning with this compound were established around the world. For example, when mercury contaminated seeds (wheat treated with a methylmercury fungicide) were used in Iraq in 1971 to prepare bread. Subsequent studies to identify the nature of such occurrences demonstrated that direct intake of water or food contaminated by methylmercury has extreme detrimental effects on the human central nervous system. Over the past three decades, contamination of the food chain by organomercury species has become of worldwide concern with increased public awareness. For example, recent studies in the Brazilian Amazon demonstrated that the Indian population had increased exposure to methylmercury because of their consumption of fish contaminated by upstream gold-mining activities. Elemental mercury used in such activities undergoes natural methylation reactions to be converted to soluble inorganic and methylmercury species: this methylmercury is able to bioaccumulate in small fish and then throughout the food chain to eventually reach the human population. Methylmercury is considered to be 50-100 times more toxic than Hg^{2+}. Thus, determination of total mercury and methylmercury are needed today.

Hair is a suitable indicator for the monitoring of human exposure to mercury and methylmercury. Also, hair reflects both recent and historical exposure to mercury (time-based exposure of 1-2 years can be monitored from individuals with long hair). Moreover, the use of hair as a mercury-control sample for possible intoxications offers practical advantageous properties (a hair sample is easier to collect and to store compared with blood or organ tissues). As a general rule, detection and final determination of mercury is carried out with the required sensitivity by CV-AAS. A previous sample pre-treatment step is recommended to clean the hair samples (e.g., washing in acetone or a non-ionic surfactant such as Triton X-100), with a final washing with redistilled water and drying at 105 °C.

For total mercury determinations, weighed clean hair samples are heated in a concentrated acid medium for sample mineralization and made up to volume in a volumetric flask before the final CV-AAS measurement.

> While the high volatility of atomic mercury is beneficial for its highly sensitive determination, it should be borne in mind that analyte losses might result during sample mineralization. This is why enclosed digestion bombs, heated by either convection or microwave irradiation, are usually employed for total mercury digestion. Alternatively, wet-digestion on a steam-bath gives satisfactory results as well, provided that the oxidation chemistry is adjusted so that the organic matrix is oxidized, and all the released mercury is converted into non-volatile Hg^{2+}.

For methylmercury determinations in hair without breaking up the methylmercury compound, one must consider that it is usually bound either to cystine sulfur or to sulphydryl groups present in other amino acids. Therefore, for the determination of the highly toxic methylmercury it is necessary to include a digestion step able to break the existing bonds between hair and the analyte, but without destroying the methylmercury compound.

Chapter Six

ELECTROTHERMAL ATOMIC ABSORPTION SPECTROMETRY

6.1. INTRODUCTION

The most successful approach in AAS, so far, to achieve detection limits in the low μg.L^{-1} level, is the use of flameless atomization, particularly electrothermal atomizers. According to IUPAC, an electrothermal atomizer is defined as "a device which is heated to the temperature required for analyte atomization by the passage of electrical current through its body". Thus, if the electrical current flows through resistance wires round the atomizer, as in the silica tube atomizer described in Chapter 5, the device should not be termed an *electrothermal atomizer* but the more general term *furnace atomizer* should be used [**Inczédy *et al* (1998)**].

Samples may be heated inside an electrothermal atomizer for final atomization of the analyte either by conductive, convective or radiative processes.

Tubular electrothermal atomizers where the atomizer itself prescribes and confines the analyte's observation volume for measurement by AAS has given rise to a very sensitive analytical technique widely used nowadays in routine laboratories: the technique is called electrothermal atomic absorption spectrometry (ETAAS). The attractiveness of this atomization device relies on features such as its high vaporization and atomization efficiency, the well-controlled thermal and chemical environment, its ability to handle high dissolved solid contents, the very low volume (microliters) of sample required (offering special benefits in those cases where the amount of sample available for analysis is limited, as in many clinical analysis), and the high sensitivity which can be achieved by ETAAS.

Since the introduction of the first commercial electrothermal atomization atomic absorption spectrometer by the Perkin Elmer company in 1970, many instrumental improvements have been developed to reduce chemical interferences and enhance accuracy, precision and sensitivity. Some of the most significant advances include the use of an autosampler to improve precision, a platform or probe atomization system to approach atomization under isothermal conditions, pyrolytically coated tubes to minimize analyte interactions within the atomizer, transversely heated tubes to allow more isothermal heating, modern background correction methods to avoid unspecific interferences, measurement of integrated absorbance and fast electronics to more accurately measure the absorbance signal (the electrothermal atomizer produces fast transient signals which are grossly distorted if the electronics are too slow to follow the rapid atomization phenomenon in the tube). A complete separate section (see section 6.6) is necessary in order to cover the enormous amount of studies carried out searching for chemicals that may act as appropriate "matrix modifiers".

Today, ETAAS is perceived by many to be a mature technique for ultratrace ($\mu g.L^{-1}$ and below) analysis at a modest cost [**Stafilov (2000)**]. Others consider it to be in a period of senescence, doomed to extinction by the proliferation of more powerful, faster and versatile (more costly, too) techniques. In any case, it is fair to acknowledge that ETAAS enjoys today a widespread use and many instruments are still sold every year. Besides, there are some areas where remarkable research with ETAAS is still going on (e.g., solid sampling, ultratrace analysis of hydride forming elements).

Finally, although it is not in the focus of this book, it has to be pointed out that electrothermal devices are also frequently employed for atomization in emission and fluorescence analytical atomic spectrometry, as well as for just vaporization in combination with other atomization/excitation or ionization sources ("combined" or "tandem sources"). In addition, the unique physico-chemical micro-environment which can be attained within the electrothermal device has also been used to advantage in the domain of fundamental atomic spectrometry and physical chemistry and examples include the determination of gas- and solid-phase diffusion coefficients of high-temperature metal vapors, sublimation heat of refractory metals, fundamental optical constants, etc [**Sturgeon (1996)**].

6.2. THE ELECTROTHERMAL ATOMIZER

Apart from high atomization efficiencies, a convenient electrothermal atomizer for analytical AAS for routine use should also meet some additional criteria such as rapid formation of the absorbing atoms, sufficiently long residence time within the observation zone, complete removal of all sample constituents after each determination, rapid cycling time, stabilization of response over time, and reasonable "large" sample volume or mass capacity to attain good sensitivity. Critical instrumental factors to be considered for optimum ETAAS performance include the length and internal diameter of the atomization cell and the material from which it is made, the precise control of the desired temperature, the heating rate, and the temperature distribution during atomization.

ETAAS was first introduced by L'vov in 1959. The system proposed by L'vov permitted sample volatilization into a well-defined volume at a high and constant temperature, selected to favor complete atomization. Unfortunately, the L'vov furnace was difficult to use for routine analysis, which lead Massmann in 1968 to introduce a simplified atomizer allowing for the direct injection of the sample onto the surface of the graphite tube (see Figure 6.1). It consists of a 5 cm long cylindrical graphite tube aligned horizontally in the optical path of the spectrometer. The tube was longitudinally heated (i.e. end-heated) by passing a high current at low voltage through the tube. This resistance heating permitted a fine gradation of the temperature (i.e. the heat program could be more strictly controlled) and the atomizer was continuously fluxed with an inert gas stream (e.g. Ar) to prevent entrance of atmospheric oxygen. The design of the first commercial instrument was based on the Massmann design. In the commercial instrument the tube was held in place between two graphite contact cylinders, which provided electrical connection. The heated graphite was protected from air oxidation by end windows and two streams of argon; an external gas flow surrounded the outside of the tube, and an internal gas flow entering between a window and the tube-end purged the inside of the tube and exited between the other tube-end and the second window. The entire assembly was mounted within an enclosed water-cooled housing.

Unfortunately, the temporal and spatial high-temperature non-isothermal conditions existing inside the tube raised a number of problems with the analysis of real samples, in particular those concerned with matrix interferences and condensation effects. It was not until 1978 when L'vov published a second, most important paper, dealing with the idea of atomization of samples from a platform placed within the tube, that a substantial reduction of matrix effects became possible. The efforts to achieve precise and accurate ETAAS analyses were decisively pushed forward with the introduction of the "stabilized temperature platform furnace" (STPF) concept of Slavin and colleagues upon which modern ETAAS is now based [**Slavin et al (1981)**]. In the STPF design several improvements were implemented to decrease interferences in the analysis by ETAAS. These advances include the use of an autosampler, a L'vov platform, an appropriate matrix modifier, Zeeman background correction, improved pyrolytically coated graphite tubes, rapid heating of the furnace, and fast electronics to detect the absorbance and integrate the signals.

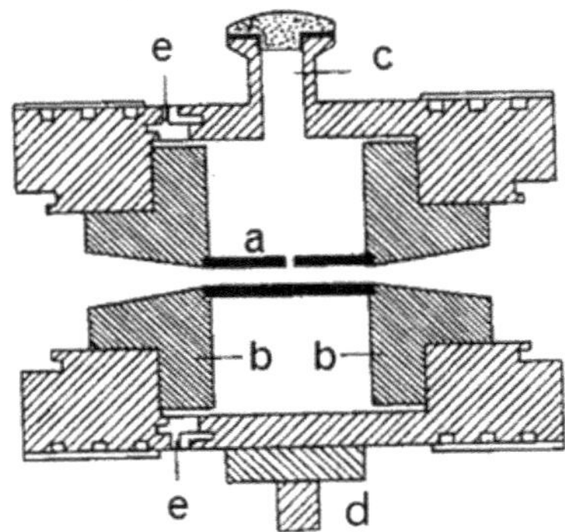

Figure 6.1. Schematics of the original Massmann design. a: graphite tube, b: steel flanges, c: sample introduction port, d: mount, e: plastic insulator.

6.2.1. The *cuvette* used in electrothermal atomic absorption spectrometry

Over the years many advances have been made to the design and materials used for the construction of *cuvettes* (or atomization tubes) for ETAAS. Of particular interest is the reduction in size as compared with early furnaces. The smaller *cuvettes* have a much reduced thermal mass and consequently achieve operating temperatures much faster than the larger *cuvettes;* besides, atoms are confined in a smaller volume. This in turn produces a higher sensitivity and greatly reduces overall atomization times leading to extended tube lifetimes. Figure 6.2 shows a photograph and a scheme of a basic *cuvette* used for analytical ETAAS. As can be seen, it consists of a cylinder (of 3-5 centimeters in length and a few millimeters diameter) with an orifice in the center of the tube wall for sample introduction. Both ends of the tube are open to allow for sample constituent removal after the analysis and transmission of the light from the hollow cathode lamp for absorption measurement.

The material used for the atomizer body must be electrically conductive and able to withstand high temperatures. In addition, good thermal shock resistance is required as well as durability and high resistance against attack by sample matrices and added chemicals. The material should also be impermeable to gaseous sample components and be able to confine the atomic vapor within the atomizer for a sufficiently long time. Other requirements are high purity, low cost and good mechanical properties.

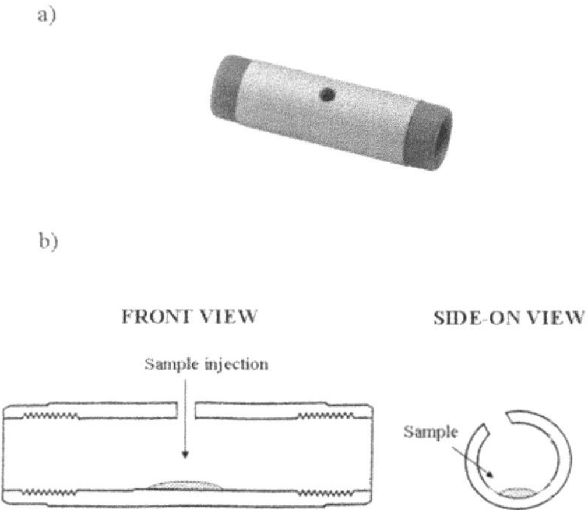

Figure 6.2. A basic cuvette for analytical ETAAS.

(a) Photograph. (b) Schematics of internal reservoir for analyte absorbing atoms.

In general, current electrothermal cells can be divided into non-graphitic (made from materials other than carbon, such as molybdenum and tungsten) and graphitic ones, depending on the substrate that constitutes the *cuvette*. Graphitic atomizers are most commonly used in commercial ETAAS instruments. The most frequent carbon material employed for the manufacturing of the electrothermal atomizers is electrographite (EG) coated with pyrographite. EG is inexpensive and consists of fine-grained isotropic graphite [**Frech (1996)**]. Unfortunately, the material is reactive, particularly at elevated temperatures and complex chemical reactions take place between sample constituents and EG during pre-treatment and atomization steps. Besides, after repeated analysis cycles, EG tubes lose substantial masses of graphite and their analytical lifetime is relatively short. Therefore, although EG tubes are easy to manufacture and competitively priced, their application is limited to volatile or medium volatile analytes, which do not react with carbon. The analytical lifetime of EG tubes is drastically increased by a pyrolytical less porous coating which provides both, a lower surface reactivity and the reduction of carbon vapor losses. Therefore, nowadays, atomizers used in commercial ETAAS instruments are most commonly made of pyrolytically graphite coated EG.

An important improvement in the design of the *cuvette* used for ETAAS was the implementation of a platform (the so-called L'vov platform) to deposit the sample within the tube (see Figure 6.3). The heating characteristics of tubes containing a platform are only slightly different from those without a platform. However, since the platform has a finite heat capacity and is heated primarily by tube radiation, its temperature will initially lag behind that of the tube wall. Thus, the sample on it will be volatilized later in time (relative to direct wall atomization), at higher tube- and gas- phase temperatures, which will favor isothermal atom formation. In other words, by using the platform the atomization time of the analyte is delayed until the tube and inert gas inside are at a more constant temperature. Thus, the platform would reduce interference effects arising from the above-mentioned temporal nonisothermal behaviour typical in tube-wall atomization.

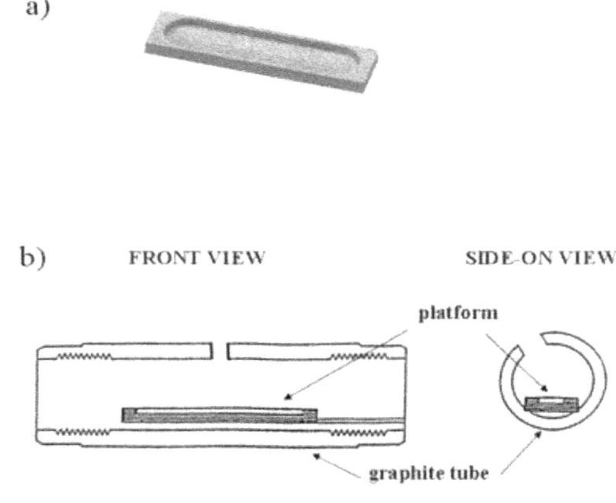

Figure 6.3. L'vov platform.

a) Photograph of the L'vov platform. As can be seen, the L'vov platform has a slight depression in the center, which can accommodate up to 50 μL of solution.

b) Schematics of the L'vov platform inserted in the graphite tube.

An alternative way proposed to achieve atomisation under isothermal conditions is to place the sample on a graphite probe. This probe is removed from the atomisation tube prior to the atomisation stage, and then re-introduced once the tube has reached the selected temperature. This system was commercialised for a while, but being much more complex than the platform and requiring an additional aperture in the tube (reducing analytical sensitivity) its fabrication and sale was discontinued.

6.2.2. Side-heated atomizers

Establishing a constant temperature requires that the tube wall be brought to the desired final temperature as rapidly as possible. Since in the conventional longitudinally heated tubes the center of the tube heats more rapidly, some time is required to establish a more or less steady temperature along the tube (that is, to achieve isothermal conditions all along at the desired atomization temperature (Figure 6.4a)). Thus, the most significant shortcoming of the Massmann-type furnace is the pronounced longitudinal temperature gradient that develops at high temperature (1000 K/cm) and accounts for the majority of the matrix interferences and condensation problems encountered using the graphite furnace as an atomizer for AAS.

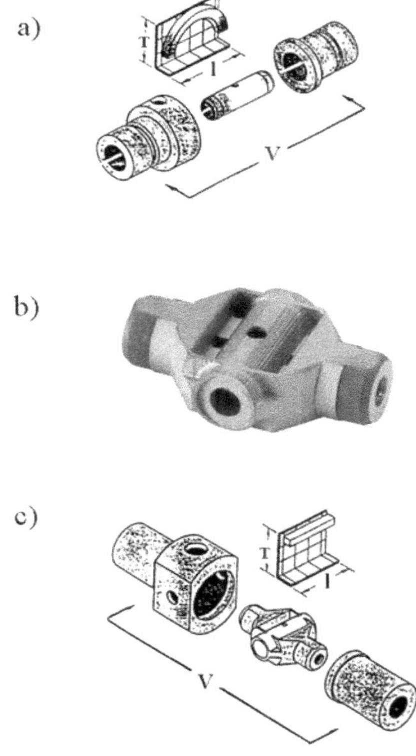

Figure 6.4. Comparison of longitudinally and transversely heated atomizer. V: voltage, l: tube length, T: temperature.

a) Scheme and heating profile of a longitudinally heated atomizer

b) Photograph of a transversely heated atomizer.

c) Scheme and heating profile of a transversely heated atomizer.

Trying to overcome such a fundamental limitation, a commercial transversely heated graphite atomizer was eventually introduced. In this case, the graphite tube (shown in Figure 6.4b) includes integral tabs, which protrude from each side. These tabs are inserted into the electrical contacts and, so, when power is applied, the tube is heated across its circumference (i.e. transversely). By applying power in this manner, the tube is heated evenly over its entire length (Figure 6.4c), thus, almost isothermal heating is achieved along the tube, significantly reducing the sample condensation problems observed with longitudinally-heated furnace systems.

An additional advantage of the transversely heated furnace is that it allows the use of longitudinal Zeeman-effect background correction. As described in Chapter 3 (section 3.6.2), longitudinal Zeeman offers all of the advantages of transverse Zeeman correction without the need to include a polarizer in the optical system, thus providing an improvement in light throughput.

6.3. STEPS FOR AN ANALYSIS BY ELECTROTHERMAL ATOMIC ABSORPTION SPECTROMETRY: THE TEMPERATURE PROGRAM

A determination by ETAAS starts by dispensing a known volume of sample into the furnace. The sample is then subjected to a multi-step temperature program by increasing the electric current through the atomizer body. As can be seen in Figure 6.5, these furnace-heating steps include: drying, pyrolysis and atomization (between the pyrolysis and the atomization process a cool down step may be included). The furnace analysis cycle ends with a higher temperature clean out step (when necessary). Then, the furnace cools down to enable the next ETAAS measurement.

Important parameters to be controlled for each step are the "final temperature" during the step, the "ramp time" or time needed to achieve that temperature and the "hold time", that is, the time spent maintaining that final temperature. In addition, the inert gas flow rate should be controlled through each step. We will now briefly consider each of these main steps:

Sample drying. This step must be accomplished in a controllable manner, such that there is a slow and even evaporation of the solvent from the matrix. In this process sample spattering should be avoided. Temperatures around 100-120°C are commonly used in this step for aqueous dissolved samples. Of course, the use of a longer ramp time provides a "gentler" increase in heating.

When a L'vov platform is used, the normal temperature lag of the platform (*versus* the tube wall) produces a natural ramping effect; therefore, shorter ramp times are usually required using platform atomization. The internal gas flow normally is left at its default maximum value (aprox. 300 mL per minute) to purge the vaporized solvent from the tube.

Pyrolysis (or ashing). The purpose of this step is to volatilize inorganic and organic matrix components, leaving the analyte in a less complex matrix for analysis. Therefore, the temperature is increased as high as possible to volatilize matrix components but below the temperature at which analyte losses would occur. Therefore, the pyrolysis temperature and the effectiveness of the step in matrix removal are limited by the temperature at which analyte atoms are lost. The internal inert gas flow is left again in this step set at approx. 300 mL per minute to drive off volatilized matrix materials (for particular sample types it may be advantageous to use an oxidizing gas, such as air, to help the in-tube sample decomposition).

With side-on heated furnaces, it is frequently an advantage to cool the atomizer prior to atomization since the heating rate is a function of the temperature range to be covered. As the temperature range is increased, the rate of heating also increases and this also extends the isothermal zone within the tube immediately.

Atomization. In this step the temperature is increased to the point where dissociation of volatilized molecular species occurs to form analyte atoms. The atomization temperature of choice will depend on the analyte. Care should be taken to avoid the use of an excessively high atomization temperature, because analyte residence time in the tube will decrease and a loss of sensitivity will occur. Also, the use of excessively high atomization temperatures can shorten the useful lifetime of the relatively expensive graphite tubes.

For appropriate atomization, it is desirable to increase the temperature as quickly as possible. In this way, the wall and atmosphere inside the tube are heated much faster than the platform, thus ensuring a stabilized tube atmosphere temperature at the time of analyte volatilization. Therefore, atomization ramp times normally will be set to minimum values in order to achieve the highest heating rate. It is also desirable to reduce or to interrupt the internal gas flow during atomization in order to increase the residence time of the atomic vapor in the atomizer.

After atomization, the furnace may be heated to still higher temperatures to burn off any sample residue, which may remain in the furnace (tube cleaning) as illustrated in the final plateau of Figure 6.5. A typical cleaning stage is 2800ºC or 2900ºC for two seconds. When refractory elements are being determined, this tube-cleaning step becomes irrelevant since the atomization temperature for those elements is very high anyway. Finally, the furnace should be cooled down to near ambient temperature prior to the introduction of the next sample.

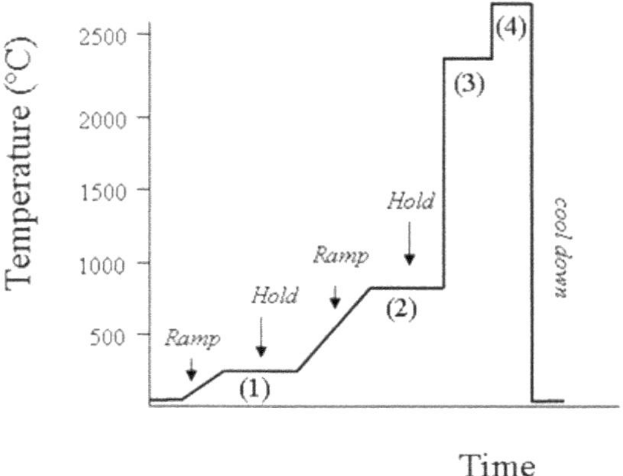

Figure 6.5. Steps of a temperature program for analysis by ETAAS. (1) Drying, (2) Ashing or pyrolysis, (3) Atomization, (4) Clean out step.

6.4. INSTRUMENTATION

The basic components of an instrument for ETAAS analysis are essentially the same as for FAAS, namely: a line source lamp, the atomizer, a background corrector, a monochromator, a light detector, and a data acquisition and data treatment system. Apart from the obvious differences in the atomizer, ETAAS instrumentation has some special requirements in some of the above components (not so crucial for FAAS analyses). For instance, particular attention should be paid for ETAAS analysis to the background corrector used as well as to the data acquisition and treatment system. Besides, it has been demonstrated that the use of an autosampler is strongly recommended in order to achieve precise enough AAS results. Finally, it should be mentioned that the optical system must provide maximum lamp light throughput while avoiding the incandescent continuum light from the inner surfaces of the heated tube.

6.4.1. Sample introduction systems: autosampler

In ETAAS just a few microliters of sample are introduced into the graphite tube. While skilled and experienced operators may obtain reasonable reproducibility by manual injection with a pipette, it is now universally accepted that autosamplers provide superior results in terms of reproducibility. Autosamplers are also used to add appropriate chemicals (e.g. "matrix modifiers" that will be discussed in section 6.6).

Another advantage brought about by the use of autosamplers is clearly that the instrument can be left working alone. This advantage is especially crucial when considering that the sample throughput in ETAAS analysis is much lower than that attained in flame AAS and so the use of the autosampler may critically reduce the manpower costs of the analysis.

6.4.2. Instrumental background correction

Instrumental methods for background correction have been already described in Chapter 3 (section 3.6). However, it has to be highlighted here that (as detailed later in section 6.5) background absorption is a more severe problem using electrothermal atomizers. During the atomization step any organic material still present in the *cuvette* is pyrolyzed and the resulting smoke may cause severe attenuation of the light beam. Besides, the presence of many salts in the *cuvette* can give rise to large background absorbance when atomized at high temperatures. All these absorptions occur also away from the atomic line wavelength and are known as "unspecific" absorption or background. For this reason, the successful development of efficient methods of background correction has been a crucial aspect for the development of a wide variety of applications and the increasing importance of ETAAS. In particular, the implementation of the Zeeman effect background correction, allowing one to correct for high sample background and structured background absorption has been a most significant advance in ETAAS analysis.

6.4.3. Data acquisition and treatment

Instrument electronics must be able to respond accurately to the fast, transient signals generated with ETAAS: as Figure 6.6 shows, as atomization begins analyte atoms are formed and the signal increases (reflecting the increasing population in the furnace), the signal will continue to increase until the rate of atom generation becomes less than the rate of atom diffusion out of the furnace. Then, the falling atom population results in a signal that decreases until all atoms are lost. Factors affecting the peak-shape characteristics are the temperature rise time of the *cuvette*, the analyte volatility, the atomization temperature, the *cuvette* material and the reactivity of the analyte with graphite or with the gas.

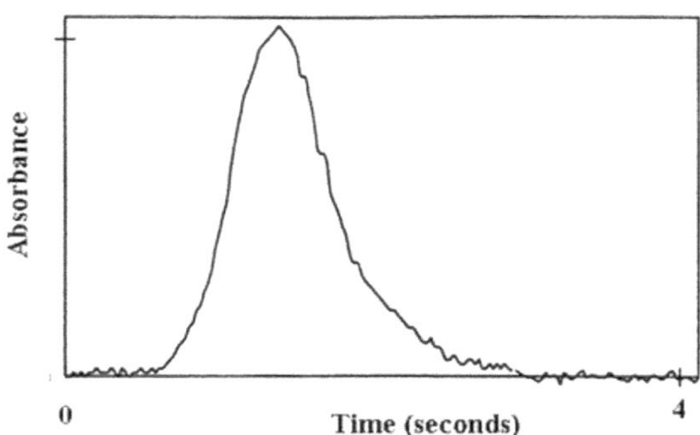

Figure 6.6. A typical absorbance peak versus time obtained with ETAAS.

The recorded "absorbance peaks" generally appear asymmetric in that they have a steep leading edge (representing the rapid atomization of analyte atoms) and a slower trailing edge resulting from diffusion of atoms from the atomizer, as Figure 6.6 illustrates, which shows a typical peak profile. As can be observed, it approximates to a combination of two different overlapping Gaussian curves, one defining the leading edge, the other the trailing edge.

To determine the analyte content of the sample, the recorded signal (peak height or peak area) must be measured. Therefore, in order to obtain good analytical results the transient absorbance signal must be measured accurately and, so, one of the potential limitations is the inadequate speed of the instrument's electronics. For many years, measuring the height of the peak was the only practical means to quantify the obtained transient signal. However, peak height measurements provide accurate signals for analyte concentrations only if: (i) there is a constant degree of atomization in the cell, (ii) the average residence time of atoms in the cell is constant, and (iii) the rate of atom formation is constant.

As matrix components can seriously affect the rate of analyte atom formation peak height measurements are seldom used today. In fact, modern instrumentation relies on the measurement of integrated peak areas: in principle, if the temperature in the atomizer is constant during the atomization step, the peak area will represent a count of all atoms present in the sample aliquot, regardless of the atomization rate. Peak area measurements, however, are most sensitive to variations produced in the long peak tail found in some analyses (see Figure 6.6). This will degrade the precision because very small changes at the end of a peak tail will become significant. It has been found that a better compromise can be to use the peak area mode, but using a reduced peak profile so that not all the peak tail is included in the final measurement (analytical signal).

6.5. INTERFERENCES

It is convenient, using electrothermal atomization, to divide potential interferences into two broad groups: (i) spectral interferences, produced by continuous emission or by light absorption either by atoms different to the analyte or, most commonly, by molecules or smoke, and (ii) non-spectral interferences, which are those affecting the production or the availability of analyte absorbing atoms.

6.5.1. Spectral interferences

Continuous emission. Emission interference arises when the intense of light emitted by the hot graphite tube or platform ("blackbody" radiation) reaches the instrument's light detector. The magnitude of this interference depends on the wavelength (visible more affected than UV lines). Emission interference is controlled basically by spectrometer optical design (in particular the slit heights), in such a way that the light reaching the detector comes from the center of the atomizer and not from the tube wall or platform, which are the sources of blackbody emission.

Background absorption. Background absorption is a non-specific attenuation of light at the analyte wavelength at the atomization step. It is caused by molecular absorption (broad band) or by light scattering caused by undissociated matrix components in the light path. Therefore, background absorption derives from an inefficient pyrolysis step.

When high background absorption is limiting the quality of the analysis, it is frequently desirable to reduce the sample size. This gives rise to a reduced mass of the background-producing matrix components with a corresponding reduction in background absorption (obviously this strategy gives rise also to a poorer analytical sensitivity). As the degree of background absorption is wavelength dependent, another alternative to minimize background absorption consists of using, when available, another absorption wavelength (background absorption is usually greater at lower wavelengths). Alternatively, using a procedure commonly known as "matrix modification" the relative volatilities of the matrix and analyte can frequently be controlled. The procedure consists of adding a reagent ("matrix modifier") to generate either increased matrix volatility or decreased analyte volatility (see section 6.6).

However, only in rare occasions is the use of the above strategies efficient enough to completely eliminate background absorption. Therefore, it is most common to compensate for residual background by resorting to instrumental approaches, described already in Chapter 3 (Section 3.6) and in section 6.4.2 of this chapter.

6.5.2. Non-spectral interferences

Non-spectral interferences in ETAAS have been widely studied. They can be due to several different processes: (i) gas-phase formation of analyte species, which can diffuse from the optical path, resulting in loss prior to dissociation into atoms, (ii) condensed-phase reaction to form an analyte compound, which is volatilised and diffused out of the absorption tube prior to atomization, (iii) co-volatilization or thermal expulsion of the analyte (in a solid, liquid or vapor phase) together with rapidly expanding matrix gases or by a carrier (or occlusion) mechanism prior to (or after) reaching the atomization temperature, and (iv) changes in the atomization path modifying analyte populations.

Strategies to minimize non-spectral interferences include the use of the L'vov platform, of "fast heating" in the atomization step, of impermeable pyrolytic-coated graphite to reduce the tendency towards carbide formation, the employment of matrix modifiers, and also the use of the standard addition method.

6.6. MATRIX MODIFIERS

If the pyrolysis step were 100 % efficient, that is, if all of the matrix could be driven off during this step, there would be no background absorption since the sample components that cause the background would be removed prior to atomization. Analyte atoms, however, must not be lost during pyrolysis. Therefore, the pyrolysis temperature and the effectiveness of the pyrolysis step in matrix removal are critical and limited by the temperature at which analyte atoms are lost [**Nóbrega *et al* (2004)**].

One of the most crucial recommendations for the chemical analysis by ETAAS is the use of an appropriate "chemical modifier" allowing a thermal separation between the analyte and the undesired concomitants during the pyrolysis stage. This added chemical modifier allows, generally, a thermal stabilization of the analyte (of special interest for highly volatile analytes or when various analyte species of different volatility, e.g., inorganic and methylated arsenic are present in the sample) and also an efficient matrix removal by volatilization during the pyrolysis stage.

The matrix modifier is typically a concentrated solution containing one or more chemical compounds that are added to the sample aliquot in the furnace via sequential deposition using an autosampler, or directly added to a solution used to dilute the sample prior to its introduction into the tube. Many different modifiers have been employed in ETAAS. The general demands of chemical modifiers include the absence of any negative influence of the modifier on the lifetime of the graphite tube, availability in high purity, low toxicity, robustness of the modifier action, etc. Also, compounds of the elements routinely determined with ETAAS (e.g. Pb, Cd, Se, As) are not desirable as modifiers. The different modifiers can be classified according to their chemical nature into several groups: (i) inorganic salts, based on ions such as Ni^{2+}, Pd^{2+}, NO_3^-, PO_4^{3-}, (ii) organometallic and complex compounds such as Ni(II)sulfonate, La(III)acetylacetonate, (iii) acids, e.g. HNO_3, H_3PO_4, (iv) bases such as NaOH, (v) oxidants such as HNO_3, MnO_4^-, (vi) reductants such as ascorbic acid, (vii) organic additives (e.g. EDTA, thiourea), (viii) gases, such as O_2 and H_2.

At present, compounds of the platinum group metals (PGMs) appear to be the most universal chemical modifiers. Combinations of PGMs with other modifiers such as $Mg(NO_3)_2$ and organic compounds (e.g., ascorbic acid, citric acid, thiourea) are also popular. Refractory carbide forming elements of the Groups 4-6 (IVA-VIA) of the periodic system (e.g., TaC, ZrC, NbC) are also frequently used [**Ortner *et al* (2002)**].

A widely pursued endeavor has been to develop a permanent chemical modifier (tube coating). Permanent chemical modification offers clear advantages such as an extended atomizer lifetime, lower reagent blanks, and the elimination of the step needed to add the modifier to each sample aliquot. The elements most commonly used for this purpose are metals with high melting points (such as iridium, rhodium or ruthenium) or a mixture of these elements with those forming very stable carbides (e.g., W, Ta, Zr). Tube impregnation is the easiest way to apply permanent modifiers to ETAAS tubes and platforms. Still more effective than impregnation is the deposition of permanent modifiers by electrolysis.

The mechanisms of action of modifiers in ETAAS are a topic of great scientific and practical interest and, frequently, the literature contains different and often contradictory proposals for mechanisms of action of modifiers and coatings. The advantageous use of PGM modifiers seems to be based on the formation of intercalation compounds and a subsequent activation of these intercalated metal atoms. These activated metal atoms are proposed to form strong covalent bonds with easily volatile analyte elements, which leads to their stabilization to remarkably high apparent temperatures. In the case of the use of refractory carbide forming elements as modifiers, it has been observed that analytes with a tendency to form oxo-anions are especially stabilized by Group 4-6 (IVA-VIA) modifiers. This can be explained by the formation of stable bronzes (mixed metal oxides with refractory metal oxides) or heteropoly-compounds acting as stabilizing analytes, which form oxoanions on the surface of carbide forming modifiers.

6.7. ATOMIZATION FROM SOLIDS AND SLURRIES

Traditionally, the determination of trace metals in solids by AAS has been carried out following a previous dissolution step. However, the direct analysis of the solid samples as they are received by the laboratory offers intrinsic advantages over conventional procedures, such as less risks of sample contamination and analyte losses, use of hazardous chemicals or pre-treatment procedures are not necessary, and less operational work is required, etc. However, there are also potential shortcomings, which will be dealt with also in this section, to be taken into account (see Figure 6.7).

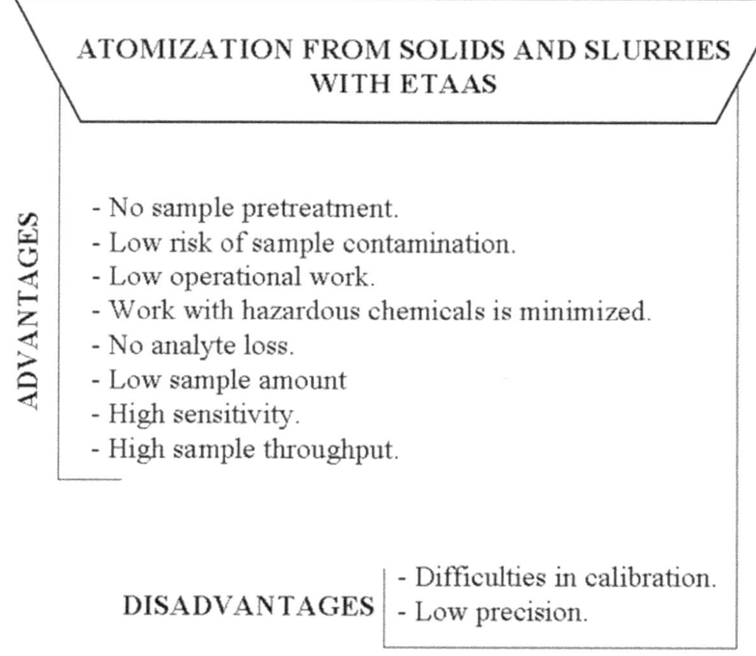

Figure 6.7. Potential advantages and shortcomings of ETAAS methods for atomization from solids and slurries.

Unlike nebulization techniques, ETAAS does no suffer significantly from particle size effects because it offers longer residence time; the availability of commercial instrumentation supporting solids analysis with the graphite furnace (e.g., accessories SSA 600 and SSA 6 from Analytik Jena AG) has led to its successful application in the analysis of a wide range of materials. Also, compared with other techniques for solid analysis, ETAAS offers high sensitivity and simplicity at moderate cost.

Solid sampling was firstly employed with a commercial atomizer in 1971. Unfortunately, some problems were associated with the direct solid sampling technique. For example, the small sample sizes required were frequently not representative enough of the sample and also matrix-matched standards were required for many applications. Slurry sampling can minimize many of these inconveniences. Slurry sampling was introduced in 1974 and it steadily grew until the 1990s, when it surpassed direct solid sampling. Nowadays, slurry sampling-ETAAS can be considered as a mature technique, which is widely utilized for metal determination in both organic and inorganic matrices, even for routine analysis [**Cal-Prieto *et al* (2002)**]. In any case, considering the advantages of direct solid sampling (e.g., best limits of detection because of the absence of any dilution and minimal risk of contamination) it should be noted that in this 21st century the use of direct solid sampling it seems is being re-visited, in particular for fast screening analyses [**Vale et al (2006)**].

In a way it can be considered that the slurry approach combines the advantages of the direct solid and liquid sampling methods: (i) while solid sampling requires one to weigh a very low mass of sample in each replicate and contamination risks during sample handling can be significant, slurry sampling allows several replicates to be obtained from just only one slurry and higher amounts of sample can be weighed, (ii) materials with high analyte content can be easily diluted using slurries, whereas the direct solid analysis requires the addition of graphite powder, which can increase the blank values, (iii) a more effective chemical modification can be carried out with slurry sampling, (iv) aqueous calibration and the standard addition method have been applied successfully for quantitative analysis, being much more troublesome for direct solid sampling than with slurry sampling, and (v) the conventional atomizers and injection systems used for liquid samples can be employed for slurries.

However, even using slurry sampling some limitations remain: probably the most critical factor is the need for maintaining the stability of the slurry until sample injection. Also, particle size may affect seriously the reproducibility and accuracy if it is non homogeneous or too large. In addition, owing to the absence of a sample pre-treatment step to eliminate the matrix, the molecular absorption signal is often rather high and structured making necessary the use of powerful background correctors. Another limitation is the likely build-up of carbonaceous residues in the atomizer (which reduce the lifetime of the atomizer). To minimize such inconveniences several experimental parameters have to be controlled for an optimum slurry analysis:

(i) *Slurry diluent.* 0.5 – 5 % nitric acid is the most common diluent used for slurry preparation. It operates as an oxidizing agent and it can be also considered as a chemical modifier. Moreover, it can efficiently increase the analyte extraction from the particles, increasing the suspension stability and improving precision.

(ii) *Stabilizing agent.* These are added to wet solid materials for dispersing agglomerates and/or to avoid particles settling down. The most frequently used are ethanol, glycerol and the non-ionic surfactant Triton® X-100.

(iii) *Particle size.* In general, particle size is a less critical factor for slurry analysis by ETAAS than for other atomic techniques for which nebulization is required and the residence time for the solid particles in the atomizer is shorter (e.g. ICP-OES). In ETAAS the most appropriate particle size for each particular situation depends on many factors including sample homogeneity and density, required accuracy and precision of the results, etc. While good results have been reported for particles as large as 300 and 150 μm diameter for relative soft materials, such as biological tissues, in general major benefits are achieved with < 50 μm diameter particles in the final slurries to be introduced in the tube.

(iv) *The mass to diluent volume ratio and the analyte partitioning***.** To establish the adequate mass and diluent volume for the slurry preparation both the sample analyte concentration and the linear analyte response range have to be considered. Concentrated slurries show a reduction in precision because of the difficulties encountered in pippeting and the higher matrix effects. The precision also worsens when using very diluted slurries owing to the statistical limitations of the technique when working with a small number of suspended particles.

The percentage of analyte extracted into the slurry liquid medium also affects the precision and accuracy: the higher the analyte extraction, the lower relative standard deviation (RSD). However, extraction is not an essential condition for accurate analysis.

(v) *Homogenization system.* A proper homogenization system of the slurry is required. Manual shaking, magnetic stirring, vortex mixing, gas bubbling and ultrasonic agitation have all been proposed as agitation modes to achieve slurry homogenization. In particular, the introduction of commercial systems with an ultrasonic probe allowing the direct slurry preparation and agitation in the autosampler cups constitutes a most positive advance in terms of versatility, operational simplicity and optimum results.

6.8. ANALYTICAL PERFORMANCE CHARACTERISTICS OF ELECTROTHERMAL ATOMIC ABSORPTION SPECTROMETRIC METHODS

In ETAAS the entire sample is vaporized and all the analyte is atomized and the atoms are confined within the tube (and so within the light path that passes through the tube) for an extended period of time. As a result, sensitivity and detection limits are significantly improved as compared with FAAS, where the burner-nebulizer system acts as a relatively inefficient sampling device. In fact, ETAAS enables the elemental determination in microliter sample volumes with detection limits typically 100 to 1000 times better than those of FAAS. Precisions achieved today, provided that sample contamination risks are carefully controlled, lie in the range of 0.5 – 5 %.

Tips to eliminate contamination risks

Due to the high sensitivity of this technique, one of the most common reasons for erroneous results in ETAAS is the contamination of samples, standards or blanks. Contamination can occur at any stage in the analytical procedure and from many diverse causes. It may arise from the reagents used in sample preparation, such as any acids needed for a dissolution step, or from dirty glassware not cleaned properly after previous use. Even the laboratory atmosphere can be a problem. To avoid contamination problems, we have listed below ten precautions, which are highly advisable in routine ETAAS work:

1. A high purity water supply, e.g., deionised water having a minimum resistivity of 10 MOhms/cm, is necessary. It is advisable to produce the water as required rather than to store it for later use.

2. Solutions of low concentration (20 $ng.mL^{-1}$ and below) should be prepared immediately before use.

3. When solutions of the same concentration of an element have to be prepared regularly it is advisable to keep the same apparatus for the same solutions.

4. All glass vessels to be used should be washed, rinsed and then soaked in 10-20 % nitric acid for at least 24 hours. Plastic vessels should be soaked in 1-5 % nitric acid. Finally, they should be thoroughly rinsed in high purity water.

5. A clean bench area should be reserved for all solution preparation, as far removed as possible from sources of dust and fumes.

6. The complete ETAAS system should, preferably, be in a clean separate room.

7. The room should be under positive air pressure, supplied from an air conditioning system with filtration against dust particles.

8. Samples must be protected from contamination by keeping sample containers sealed until they are to be processed, and minimizing the time they are open to the atmosphere.

9. For many analyses, acid dissolution is necessary for sample pre-treatment. Contamination introduced from the acid used can be serious, even when using analytical grade reagents and, therefore, high purity reagents should be used.

10.Last but not least, sample handling does not start when the sample reaches the laboratory. Special care has to be taken in the sample collection and in all the steps undertaken prior to its arrival to the laboratory (analysts and client communication is critical).

It has to be noted that the magnitude of the ETAAS signal observed depends on analyte mass rather than on concentration. Therefore, it is also common to express detection limits in mass units (pg). This means that increasing the sample volume inserted in the graphite tube can lower detection limits. However, the maximum sample size that can be accommodated in the tube is limited in real practice and should be kept in mind. Maximum volume that may be pipetted onto a L'vov platform is 50 μL and if a large concentration of nitric acid or a surfactant are added, that maximum sample volume must be reduced. Due to likely spreading of the sample using too large a sample volume may result in the spilling of a drop(s) out from the platform.

With regard to interferences: while spectral interferences are uncommon, non-spectral interferences are more serious here and they demand adequate instrumentation (e.g., proper graphite tube, proper background corrector) and also an appropriate chemical modifier in order to minimize them.

Being based on atomic absorption, ETAAS is basically a single-element technique. However, ETAAS analysis times are longer than those for flame sampling, mainly due to the need to thermally program the system to remove solvent and matrix components prior to atomization. Therefore, ETAAS has a comparatively low sample throughput: a typical determination by ETAAS normally requires 2-4 minutes per element for each sample. Fortunately, as a counterpart, the ETAAS technique does not need attended operation if sample introduction is accomplished with an autosampler. Thus, it can be kept running continuously all day.

6.9. APPLICATIONS AND CASE STUDIES

The enhanced sensitivity of ETAAS and its ability to analyze very small samples significantly expands the capabilities of atomic absorption. The technique has proven particularly attractive for the analysis of toxic and essential ultratrace elements in clinical samples, where a limited amount of sample is usually available. Also, elemental analysis of biological specimens can be carried out with good accuracy and precision after fast pretreatment within the atomizer during the pyrolysis step, thus avoiding unnecessary digestion or deproteinization procedures (and, thereby, minimizing sample manipulation and contamination risks).

ETAAS is also routinely used in many other applications where ultratrace elemental analysis is required. Food analysis, environmental control, ultratrace element impurities in high purity materials, etc, are other important fields of potential applications. As pointed out before, the low limits of detection of the technique demand considerable care to avoid sample contamination. For that reason it is recommended to carry out sample pre-treatments and analysis in a clean room, all the reagents should be of maximum purity, containers should be especially washed, etc.

In order to illustrate typical applications of ETAAS used today, the determination of a toxic metal (lead in biological fluids), of an essential element (selenium) in the main food source for newborns (breast milk), and of arsenic in sediments and soils, have been selected and described below.

6.9.1. Determination of lead in human urine and blood

Lead is a non-essential, toxic element that has been shown to be particularly harmful to young children (in fact, lead poisoning has been described as the number one environmental disease affecting children in the U.S.). The recommended clinical test for assessing lead exposure is the determination of lead in whole blood and urine. Although many different approaches have been proposed for lead determination in human urine and blood by ETAAS, probably the most popular one is based on the dilution of whole blood with a phosphate modifier, followed by L'vov platform atomization and integrated absorbance measurements. Phosphate is a "hard base" (according to the Pearson classification), and tends to bind those analyte ions that are "hard" or "borderline acids" such as Pb, Ag, Cd, etc, forming precipitates that are transformed during thermal pre-treatment into pyrophosphates and other intermediate species. The procedure is very simple because samples are just diluted with a solution containing the surfactant Triton® X-100 (surfactants reduce the surface tension of liquid samples, providing better contact, larger surface coverage or just better dilution/dispersive effects in the sample), nitric acid (which facilitates ashing of the sample) and the phosphate matrix modifier. In particular, $NH_4H_2PO_4$ is widely used as a modifier [**Parsons *et al* (1993)**], thus combining the beneficial effects of both phosphate and ammonium moieties (the ammonium combines with the chloride ions from the sample forming the highly volatile ammonium chloride, which is eliminated during ashing).

In recent years a lot of effort has been also devoted to investigations of using permanent modifiers for lead analysis. Permanent modifiers successfully tested include Ir, Zr+Ir, W+Ir, W+Rh, Zr+Rh or Ir+Rh. For example, for the case of iridium up to 1100 firings were possible with the same coating without sensitivity losses [**Grinberg et al (2001)**]. The samples were treated with 0.1 % Triton® X-100 and 0.2 % nitric acid and the selected pyrolysis temperature for both matrices was 800 ^{o}C.

6.9.2. Determination of selenium in human milk

Selenium is an essential trace element for humans, as a component of two enzymes, glutathione peroxidase that protects against oxidative damage and deiodinase (deiodinases can either activate or inactivate thyroid hormones). It has also been identified in human plasma as part of the selenoprotein P. Low selenium intake has been associated with the Keshan disease, a cardiomyopathy in children, and the Kashin-Beck disease in children and teenagers. Newborn babies, particularly those of low birth weight may be at greater risk of selenium deficiency. However, presence of this element seems to require a rather narrow essential range because the toxic doses are only about 100 times higher than those required for normal function, consequently the accurate and sensitive determination of selenium in foods – including, of course, human milk – is of particular importance.

Selenium is considered as a volatile element and its ETAAS determination is affected by losses during ashing, so thermal stabilization by chemical modification is crucial. Other difficulties include the fact that in the complex milk matrix, selenium may exist in a wide variety of chemical species. Moreover, the resonance wavelength of 196.0 nm is in the far-UV region where molecular absorption can cause serious problems. In particular, spectral interferences caused by the thermal decomposition of iron and phosphorus compounds are difficult to correct for with the deuterium lamp, it being advisable to resort to the Zeeman background corrector when analysing serum, urine, etc. Although in some reports interference-free conditions have been claimed for Se, the majority of published methods use matrix-matched standards, i.e., by adding standards to a sample without detectable selenium or using a synthetic solution with the corresponding inorganic composition as the blank.

In food analysis, several metals have been used as chemical modifiers for selenium determination by ETAAS, such as nickel nitrate and palladium nitrate. Also, the combination of a palladium modifier with ascorbic acid as a reducing agent has recently demonstrated its utility for the analysis of total selenium in samples such as wheat flour [**Alexiu *et al* (2005)**]. Also, it has been demonstrated that ammonium molybdate reduces the phosphorus spectral interference when measuring selenium by ETAAS with conventional deuterium background correction, using mussels and clams as "model samples".

For the particular case of human milk, an ETAAS method was developed based on the sample digestion with a HNO_3 + H_2O_2 mixture in a microwave oven and determination using a mixed modifier Zr-Ir (the modifier solutions were prepared by dissolving the appropriate amounts of the salts $ZrOCl_2.8H_2O$ and $Na_2IrCl_6.6H_2O$). The method was applied to the analysis of breast milk of Greek women. Results showed that selenium content was in the range 16.7 – 42.6 μg L^{-1} [**Theodorolea *et al* (2005)**].

6.9.3. Determination of arsenic in sediment and soil slurries

A large number of investigations dealing with elemental analysis of biological tissues and inorganic materials have demonstrated the analytical benefits of simplifying sample preparation procedures by using slurry sampling combined with ETAAS. For the analysis of arsenic in sediments and soils a comparison has been made between the use of a permanent W+Rh modifier with the conventional $Pd+Mg(NO_3)_2$.

Slurries were prepared by accurately weighing 0.1-0.5 g of the sample and final dilution to the required volume with 0.04% Triton® X-100 containing

0.5% v/v HNO_3. The resulting slurry was homogenized in an ultrasonic bath for 10 min in order to break up particle agglomerates. Finally, prior to pipetting into the atomizer, the slurry was homogenized again by sonication at about 10 W for 20 s [**Barbosa *et al* (2000)**]. The results, in terms of precision and accuracy of the method, employing the W+Rh platform treatment for this arsenic determination, were always better than those achieved using the conventional Pd+$Mg(NO_3)_2$ modifier, or even better than those obtained using a conventional digestion of the corresponding solid samples.

Chapter Seven

FLOW ANALYSIS AND ATOMIC ABSORPTION SPECTROMETRY

7.1. INTRODUCTION

Flow-based methods based on microfluidic manipulation of samples and reagents are widely used nowadays to carry out the many varied sample pre-treatments, on-line with the detection system. These methods rely on three aspects which should be kept reproducible: (i) volume of sample injected into the flow manifold, (ii) dispersion of the injected sample throughout the conduits of the manifold (precise pumping is required), and (iii) timing for each sequence carried out in the flow system (i.e. mixing with other streams, time to reach the detector, etc). In this way, every sample and standard is subjected to identical treatment and measured after the same time interval, so there is no need to wait for complete reactions. The use of sample manipulation flow-based methods brings about several important advantages, as compared to manual batch procedures, both in terms of enhanced sample manipulation and of analytical performance. The advantages are collectively shown in Figure 7.1.

Automated sample processing can be based either on continuous flow or on programmable flow. Continuous flow techniques, such as flow injection analysis (FIA), use constant forward motion of the carrier to transport the sample from the injector to the detection system. Ruzicka coined the name of FIA in 1975, and the first reports on the methodologies explicitly defined as FIA for atomic absorption spectrometry were published in 1979 [**Wolf *et al* (1979), Yoza *et al* (1979)**]. In FIA techniques, samples are injected into a carrier solution which transports the sample zone to the detection system, while desired (bio)chemical reactions take place (typically, the analyte carrier stream and a stream of reagent are mixed at a confluence point). The mixed stream flows constantly through the detector giving rise to a transient signal.

As can be seen in Figure 7.2a, FIA designs allow for a high transparency of operation, because the on-line processes are spatially separated.

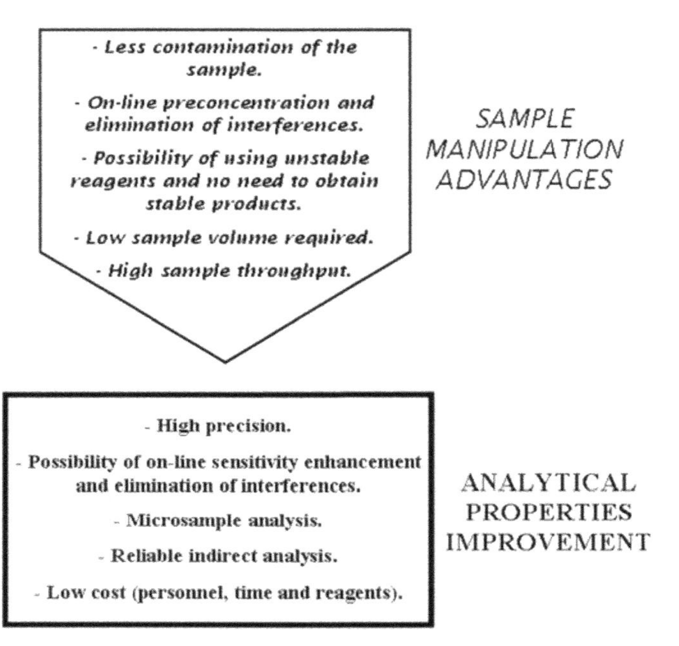

Figure 7.1. Advantages of FIA systems in comparison with batch sample pre-treatment procedures.

> In FIA analysis the profile "concentration versus time" (known as the fiagram) observed at the detector, for an injected sample with an analyte concentration C_o is a peak. This peak is the result of the dispersion processes occurring to the sample in the flow. The ratio of the injected analyte concentration, C_o, to the instantaneous concentration corresponding to any point of the sample volume reaching the detector, C_g, is known as the dispersion coefficient, D_g. The value of the dispersion coefficient at the peak maximum, D, can be used to characterize the extent of mixing in a given FIA manifold. In most FIA systems, D values are higher than unity. However, whenever a preconcentration step has taken place, the resulting D values are lower than unity.

Programmable flow techniques, such as sequential injection analysis (SIA), use flow reversals and flow acceleration to mix sample with reagents. SIA techniques were first developed in 1990 [**Ruzicka *et al* (1990), Lenehan *et al***

(2002)]. The heart of a SIA manifold is a multi-position selection valve (Figure 7.2b). Fluids are manipulated within the manifold by means of a bi-directional pump and accurate handling of sample and reagent zones requires computer control. A holding coil is placed between the pump and the common port of the multi-position selection valve. The selection ports of the valve are coupled to sample and reagent reservoirs as well as to a detector. In a first step the valve is directed to the selection port that is connected to the sample line and a zone of the sample is drawn into the holding coil by the pump. Then, the selection valve is directed to a port that is connected to a reagent line and a zone of the reagent is drawn up into the holding coil adjacent to the sample zone. Finally, the selection valve is switched to a port that is connected to a detector. As the zones move through the reaction coil towards the detector system, zone dispersion and overlap occurs.

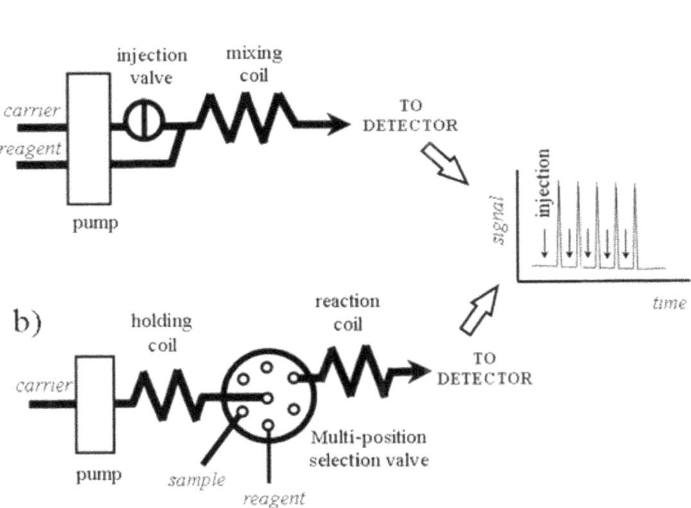

Figure 7.2. Schematics of flow injection and sequential injection manifolds. a) Basic two-channel FIA manifold. b)Simple SIA manifold.

The majority of SIA procedures are based on solution-phase chemistry, although in recent years [**Economou (2005)**] the scope of SIA has been extended to more complex, on-line sample-manipulation processes. Notwithstanding, considering the broader applications and the easier "visualization" of FIA procedures, predominant attention will be devoted to FIA-based methods in this book.

The on-line coupling of chromatographic separations (flow systems) to atomic absorption spectrometry is also straightforward. The information given by such hybrids is unique, in particular to solve metal speciation problems. Therefore, Section 7.9 discusses the coupling of chromatographic techniques to atomic absorption spectrometers.

In order to highlight the analytical potential of these combined techniques, this chapter ends by describing three selected analytical problems and their resolution by resorting to flow analysis-AAS.

7.2. FLOW INJECTION ANALYSIS AND ATOMIC ABSORPTION SPECTROMETRY

As shown in Figure 7.3, the variety of sample manipulation processes that can be carried out by flow operation procedures coupled to atomic detection covers a wide scope. These processes can be classified according to a hierarchy going from simple operations such as on-line dilution or standard additions, to preconcentration/separation stages using two-phase systems, sample digestion and even to *in situ* uptake of samples. Although a higher degree of manifold complexity is expected the larger the applications square in Figure 7.3, it is clear that flow injection systems offer high versatility, allowing the opening of new automation paths to facilitate routine analysis by AAS.

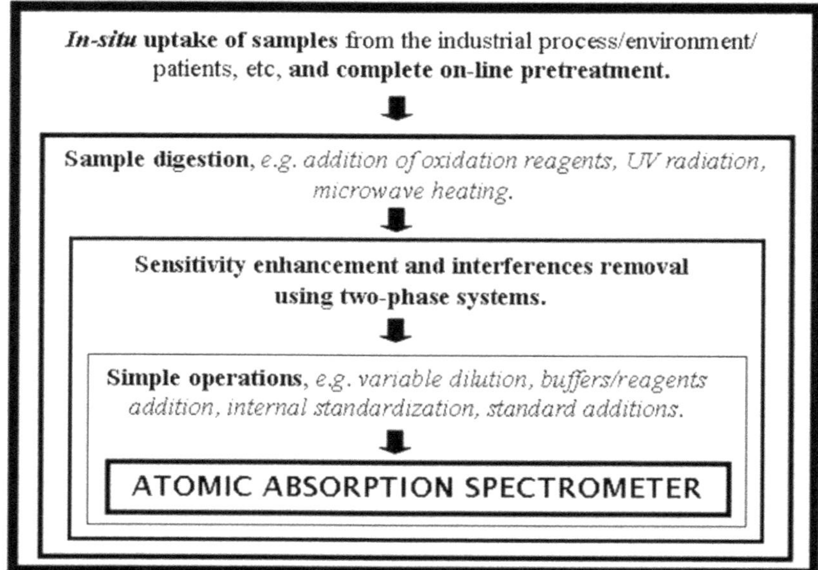

Figure 7.3. Scope of the sample pre-treatment processes covered by on-line flow strategies.

Processes like dilution, reagents mixing or standard additions can be considered as the simpler pre-treatment operations and robust on-line flow manifolds to enable these sample pre-treatment stages are now routinely used. Moreover, flow systems can easily be integrated with non-chromatographic separation/preconcentration techniques on-line with the detector. Such separation/preconcentration techniques are based on the use of two phases (e.g. solid-liquid, gas-liquid or liquid-liquid) to obtain two sample fractions, one of them containing the enriched analyte(s), free from potential matrix interferences, and the other one containing the matrix.

On-line flow systems for the decomposition/dissolution of solid samples constitute a further extension of sample pre-treatment procedures enhanced by the use of flow manifolds. In this development, different approaches including on-line chemical oxidation, photo-oxidation and microwave heating are now successfully exploited. Finally, the *in situ* uptake of samples straightforwardly from their source and their complete on-line pre-treatment before atomic detection can be considered as the ideal to be pursued in terms of automation. This goal has been already addressed and interesting manifolds have been proposed to wholly automate AAS analytical measurements in clinical laboratories (e.g., the procedure starts with the *in vivo* sample uptake of blood samples), for *in situ* environmental analysis, and also for industrial process control.

7.3. INSTRUMENTATION

The basic units of the more common flow manifolds are described in this section. Examples of integrated flow systems are detailed from section 7.4 to 7.7.

The three basic components of a flow system are the sample injection port, the propulsion unit and the connecting tubes. Special components, allowing for on-line separation/preconcentration, are also often introduced into the basic manifold. While the instrumentation used for such systems will be detailed in sections 7.5 and 7.6, specific components of manifolds for sample digestion are dealt with in section 7.7.

7.3.1. Sample introduction unit

An injection port inserted into the flow system is used to introduce samples into the FIA manifold. The injection port should allow one to precisely introduce a volume of sample as a plug into the continuously moving carrier stream in such a way that the movement of the stream is not disturbed. In the early stages of FIA development, the liquid samples were manually introduced with a syringe through a rubber septum. However, this was far from ideal since lack of reproducibility in the manual injection, caused by components leaching from the septum material and leakage after repeated injections at the same spot, was among the problems occurring with such a sample introduction unit.

Since then, different sample introduction approaches have been investigated, for example, *inter alia,* valveless injection systems where the sample introduction is time controlled and volume-controlled sample introduction systems. Figure 7.4 shows the schematics of a commonly used six-way sliding rotary valve. This commercially available volume-based injection port has an external loop. In the "load" position the loop is filled with the sample, while in the "injection" position the content of the loop is dragged through the manifold by the carrier solution. To introduce a different sample volume the external loop has to be replaced.

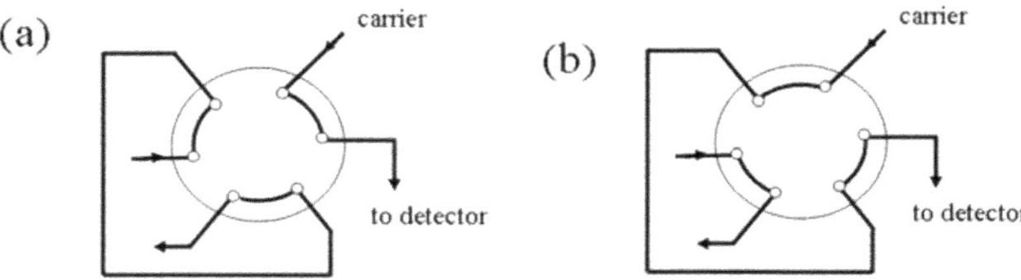

Figure 7.4. Diagram of a six-way rotary valve. a) Load position. b) Inject position.

7.3.2. Propulsion system

A propulsion unit should provide a continuous and reproducible flow rate of carrier solutions through the manifold. Given that in some cases more than one solution has to be propelled, multichannel capabilities are advisable. In addition, the ideal delivery system should allow one to select the appropriate flow rate and provide a pulse-free flow. Three main types of propulsion strategies have been proposed for the combination of flowing systems with atomic detectors:

i) a propulsion-less manifold, which relies on the negative pressure generated by the nebuliser to draw the liquid solution into the instrument.

ii) the use of a gas-pressurized carrier reservoir to propel solutions through the manifold. The gases commonly used are air or nitrogen. Although this system gives rise to a pulse-free flow, drawbacks such as the consumption of gases or the difficulty of achieving constant flow rates have greatly limited its use.

iii) the most commonly used propulsion systems by far are electrical pumps, such as peristaltic pumps and syringe pumps.

Figure 7.5 shows a schematic of the pumping principle of a peristaltic pump. The flow rate depends on the rotation speed and the inner diameter of the pumping tubes. Nowadays, a wide variety of flexible tubes with different inner diameters and made of different materials (e.g., Tygon®, poly(vinyl chloride) (PVC), PVC derivatives, silicone rubber) are commercially available, the choice being dependent upon the solution to be pumped (e.g., acids, organic solvents). Peristaltic pumps tend to give slight flow pulses (giving rise to non stable AAS signals), which can be greatly reduced by an appropriate adjustment of the clamps pressing the pumping tubes.

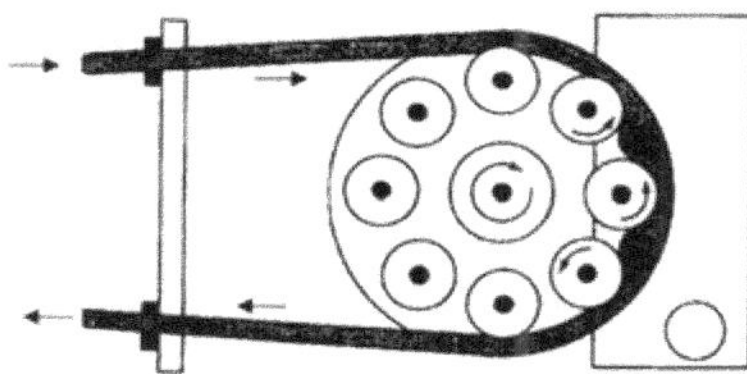

Figure 7.5. Head of a peristaltic pump.

The syringe pump is a large, electrically operated simulation of a hypodermic syringe. Syringe pumps can be used even if the backpressure is relatively moderate (for example, backpressure due to the incorporation of a capillary or a packed minicolumn in the flow system). A syringe pump exhibits higher precision than a peristaltic pump when a microflow delivery is required.

7.3.3. Connecting tubes

Inert and flexible tubing with inner diameters in the range 0.3–1.5 mm allows the connection of the different units of the flow manifold. In some parts of the system, the tubing is bent with a coiled-shape to allow for an efficient mixing of the reagents and the sample. Multiway connectors are frequently used to mix flows from different channels.

7.4. SIMPLE COMMON MANIFOLDS: DILUTION AND CALIBRATION

Many types of manifolds can be mounted using the components described in the previous section, and some are commercially available as accessories for atomic absorption spectrometers. To illustrate such versatility, Figure 7.6 shows several diagrams of very simple flow manifolds. The manifold shown in Figure 7.6a is the simplest one (single-line manifold). Apart from the obvious advantages of using this system (e.g., analysis of small sample volumes, high sample throughput), its combination with AAS allows the straightforward analysis of samples with high viscosity or with important dissolved solids content, since blockage of the nebuliser or the burner is minimised. The decrease of sample volume or an increase of the mixing coil dimensions will give rise to higher dispersion of the sample injected and therefore lead to higher dilution of the sample. The single-line manifold has been proposed for standard addition using a configuration called "reagent-injection FIA". In such a configuration, the sample is used as the carrier stream into which standards (or reagents) are injected.

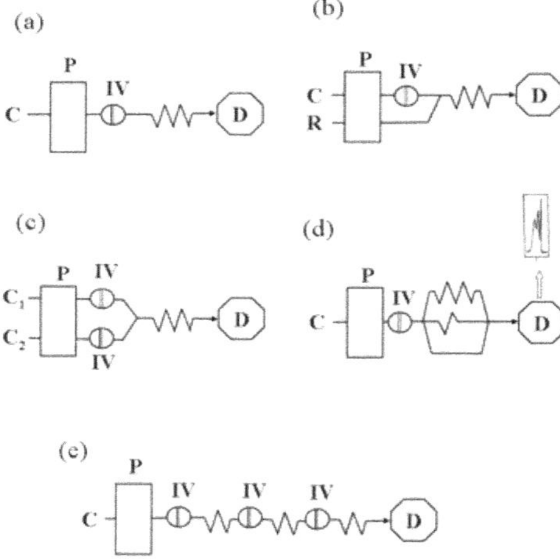

Figure 7.6. Selected examples of simple flow operation systems. P, peristaltic pump; IV, injection valve; C, carrier; R, reagent; D, detector. (a) Single line; (b) multi line; (c) merging zones; (d) three-branch on-line sample dilution system; (e) simultaneous sample injections flow system.

The addition of a second channel (Figure 7.6b) allows one to carry out some manipulations or pre-treatment of the sample, such as dilution, addition of reagents and of masking agents (e.g., addition of caesium or lanthanum in flame atomic absorption spectrometry). The flow system depicted in Figure 7.6c, called "merging-zones", allows for important savings in reagent consumption as compared to the system shown in Figure 6.b configuration and is also of particular interest for on-line standard additions.

As it was pointed out, the passage of the sample-dispersed zone through to the detector produces a FIA transient signal proportional to the analyte concentration in the sample, usually recorded as a peak. For most applications, peak heights (associated with the most concentrated portion of the dispersed zone) or peak areas are measured. However, measurements related to other portions of the dispersed zone, where the concentration gradients are more pronounced, are also of analytical interest. Exploitation of gradients has expanded the range of applications of FIA, allowing calibrations based on a single standard solution, selectivity evaluations, etc. However, the precision of measurements may deteriorate with fluctuations in flow parameters, the effect being more severe when the measurements are carried out in a region of the dispersed zone with pronounced concentration gradients. The manifolds shown in Figures 7.6d and 7.6e have been proposed to overcome such effects.

The set-up depicted in Figure 7.6d allows one to achieve a variety of dilution factors with a single injection. The manifold consists of the use of tubing with different lengths, such that the residence time of the sub-sampled zones in each branch are different. Multiple peaks are formed on recombination. In general, for *n* unequal branches, *n* peaks are produced. For example, Figure 7.6d illustrates a three-branch manifold, which could give rise to three partially overlapped peaks (with five possible measurement points, three maximum and two minimum). The method generates a family of calibration graphs of varying sensitivity.

In the manifold of Figure 7.6e the required measurements are made in portions of the dispersed zone in which the concentration gradients are minimized. To operate the system, several plugs of a given solution are simultaneously inserted into the same carrier stream, allowing them to overlap strongly as a consequence of dispersion. Provided that the plugs of the same solution are used, their asynchronous merging results in a final dispersed zone characterised by several sites without appreciable concentration gradients, corresponding to the maximum and minimum values of the concentration/time profile. The measurements related to these sites offer a high reproducibility because they are less affected by fluctuations of the flow parameters.

7.5. SOLID-LIQUID SEPARATION AND PRECONCENTRATION

Solid-liquid separation manifolds may be classified according to their separation principles and the medium used for retention: (i) sorption on a solid phase, generally packed into a minicolumn inserted in the flow system, (ii) precipitation and co-precipitation, based on a combination in a flow of precipitation of the analyte, its filtration and final dissolution. In comparison with precipitation methods, co-precipitation is less demanding in terms of the solubility of the precipitate formed.

7.5.1. Sorption

Many sorbents have been developed so far and some of them are marketed as pre-packed minicolumns. The characteristic features of on-line preconcentration flow systems demand particular properties of the packed material, which may be only of minor importance in batch or traditional column procedures: in flowing systems the solid-phase has to be reusable and to offer high mechanical resistance, also the kinetic processes or reactions have to be rapid (analytes have to be efficiently retained and readily eluted). Besides, in order to avoid backpressure and non-uniform flow patterns it is convenient that the packing material does not swell or shrink when passing the required solvents (e.g., carrier and eluent).

Sorption preconcentration methods reported can be divided into two general groups (see Figure 7.7):

(a) analyte ions are collected directly by a selective solid-phase and then eluted from the solid with an appropriate eluent. The sorption of ions involves the use of solid-phases (containing a suitable ion exchanger, a chemical reagent, or even microorganisms) immobilized on the solid support packed in a minicolumn The solid active phases can be purchased (e.g. Chelex® 100 which consists of iminodiacetate groups on a styrene divinylbenzene support, or 8-hydroxiquinoline bound to controlled pore glass), or prepared by the working laboratory. An extensive review containing more than 250 references about solid-phase extraction of trace elements has been published by **Carmel (2003)**. Eluents commonly used are acids, bases or complexing agents. To show the practical utility of this approach, section 7.10 discusses the sensitive determination of aluminium in dialysis concentrates by using an on-line minicolumn.

(b) metal ions are adsorbed on suitable solids (e.g. activated carbon, octadecyl functional groups bonded onto silica gel, etc) as hydrophobic metal chelates (formed on-line in the flow system), to be later eluted with a hydroorganic solvent. The most commonly used chelating reagents are dithiocarbamate derivatives possessing very active sulphur metal-binding sites.

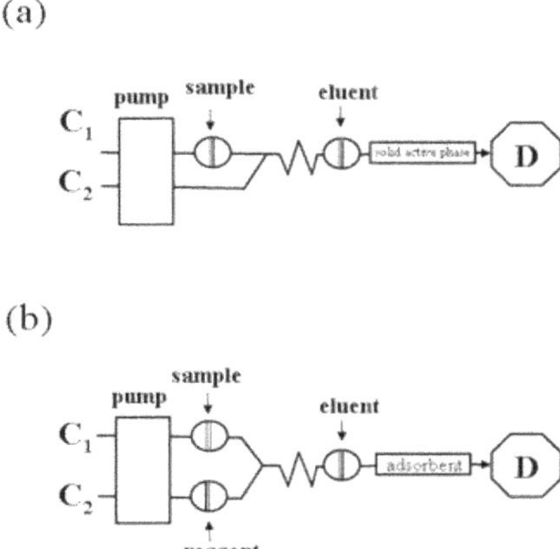

Figure 7.7. Basic manifolds for on-line preconcentration of a solid phase. a) System for sorption of ions. b) System for flow metal-chelate formation and subsequent sorption.

This latter approach is of particular interest when reagents with slow binding kinetics to the metal analyte are used. In addition, it is appropriate when the bond between the metal and the reagent is too strong, since there is no need to break down the bond during elution. However, it has some disadvantages, e.g., the sorbent support can become saturated with the free reagent (and therefore the metal chelate retention is decreased). Furthermore, the hydroorganic character of the solvents used as eluents can cause problems with some atomic detectors.

7.5.2. Precipitation and co-precipitation

Continuous precipitation – filtration - dissolution processes can be considered as continuous solid-liquid systems in which the second phase is first generated and then disappears *in situ*. The physical chemistry involved in precipitation and co-precipitation flow systems has some resemblance to metal-chelate sorption (a filter replacing the sorbent column), except that no column capacity limitations exist for precipitation/co-precipitation methodologies.

7.6. GAS-PHASE FORMATION STRATEGIES

Mass transfer between a liquid initially containing the analyte and a gas phase, which becomes enriched by the analyte (or a gaseous derivative), offers some interesting features in combination with atomic detectors. In such systems, 100 % of the analyte in gaseous form is introduced into the atomic detector (as compared to 1-5 % of the analyte introduced by conventional liquid nebulisation) giving rise to important sensitivity improvements. In addition, as already explained in Chapter 5, a striking suppression of interferences during atomisation is also achieved.

7.6.1. Flow systems for the formation of volatile derivatives of the analyte(s)

The most popular volatile compounds used in combination with AAS are covalent binary hydrides and mercury cold vapour. Despite its well known advantages, classical batch mercury cold vapour or hydride generation procedures have a number of pitfalls which can be overcome with the use of flow systems, including: (i) relatively modest throughputs; (ii) poor precisions due to incomplete control on reaction conditions, particularly reaction time; (iii) the common hydride generation process used also generates hydrogen in excess which, if released as a sudden burst, could affect the flame; (iv) quite large amounts of sample are needed. Such drawbacks can be reduced by the use of flow procedures. However, on the other hand, such use of flow systems will require rapid analyte volatilization kinetics (e.g. fast chemical reactions) and efficient mass transfer between both phases (the co-occurrence of the two phases in a hydrodynamic system gives rise to some problems that call for ingenious technical solutions).

The earliest continuous hydride generation system connected on-line to AAS was reported by [**(Astrom (1982)]**. Since then, on-line methods for mercury cold vapour and covalent hydride generation have become well established. More recently, among other analytically useful volatile compounds, it should be highlighted the great analytical potential demonstrated by the generation of cadmium cold vapour [**Sanz-Medel *et al* (1995)**] and alkyl derivatives of several metal(loids) [**Rapsomanikis (1994)**].

In continuous hydride or cold vapour generation systems the sample is mixed on-line with the reducing agent. The volatile species formed are separated from the liquid in a gas-liquid separator (GLS) with the aid of a continuous flow of a stripping gas. The stripping gas and the volatile species (analyte compounds + by-products continuously produced such as H_2, CO_2 or water vapour) are continuously swept to the detector, while the liquid is driven to waste. Argon, at a constant flow rate generally between 50-500 mL/min^{-1}, is most frequently used as the stripping gas.

In addition to the factors affecting the sensitivity of batch hydride or cold vapour generation methods (e.g., matrix components, concentration of reagents and type of detector) the following operational parameters have to be also considered for continuous systems: the flow rate of the sample channel, the ratio of flow rates sample/reducing agent, the flow rate of the stripping gas and the design of the GLS. Separators to be used in continuous systems should meet two principal requirements: (i) they should work smoothly and regularly in order to avoid both irreproducibility and to decrease signal fluctuations, and (ii) they should induce minimal dispersion or dilution of the analyte in gaseous form, i.e. they should have dead volumes as low as possible, though not so low as to give rise to incomplete gas-liquid separation.

Many different designs have been evaluated as GLSs, some of them are shown in Figure 7.8 (not drawn to the same scale). The most common designs consist of chambers made of glass. The use of gas permeable membranes (in connection with flat or concentric hollow cylinders) has also been proposed to separate gases from liquids in flow systems; for example, some practitioners have utilized a microporous poly(tetrafluoroethylene) (PTFE) membrane as a diffusion medium for the separation of the gaseous analyte derivative from solution. Figure 7.9 depicts a typical FIA manifold to achieve on-line gas-liquid separation.

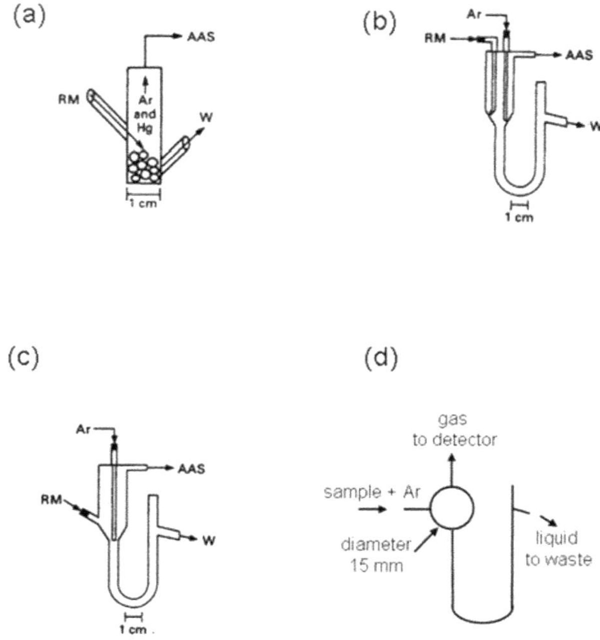

Figure 7.8. Examples of gas-liquid separators used in flow systems (not drawn to same scale). RM: reagent mixture containing the analyte; W: waste. a, b and c from C.P. Hanna, P.E. Haigh, J.F. Tyson, S. McIntosh, J. Anal. At. Spectrom., 8 (1993) 585. Reproduced with permission of the Royal Society of Chemistry. d from Y.C.C. Chan, Anal. Chem. 57 (1985) 1482. Reproduced with permission of the Americal Chemical Society.

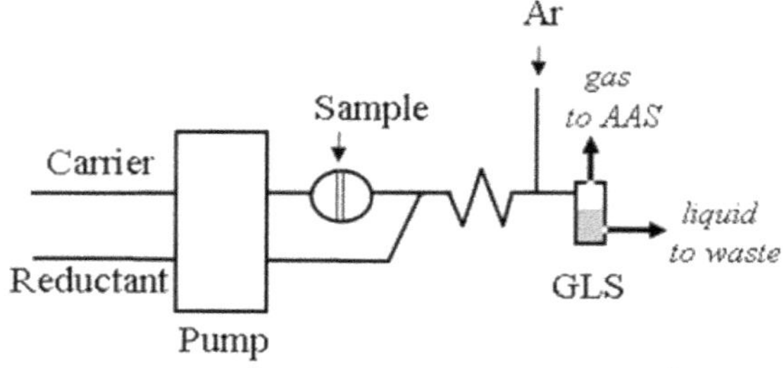

Figure 7.9. Basic FIA system for on-line gas-liquid separation.

Nowadays most manufacturers of atomic absorption spectrometers market also flow injection systems for cold vapour/hydride generation, as an accessory. Furthermore, compact atomic absorption spectrometers specifically dedicated to the fully automated and fast determination of mercury using a flow system are also available.

7.6.2. Approaches for preconcentration in the gas phase

Higher increases in sensitivity can be obtained by further concentration of the analytes released into the gas-phase. In this context, the amalgamation of atomic mercury with noble metals and later release by heating is a well-established method (commercial equipment is available) to further concentrate mercury vapour. The trap usually consists of gold-coated silica powder, gold-coated sand, or amalgams such as gold-platinum gauze.

Approaches to trap the hydrides involve the use of cryogenic traps or the retention of volatile hydrides on a slightly heated surface (normally graphite coated with noble metals) followed in both cases by release by further heating. Such approaches are most frequently used in connection with detection by ETAAS and so they will be explained in more detail in section 7.8.

The sensitivities obtained with the above-mentioned gas-phase preconcentration methods are very good and primarily limited by their blank values.

7.7. SAMPLE DIGESTION

The possibility to carry out the decomposition/dissolution of solid samples on-line with atomic absorption spectrometers has attracted a great deal of interest. Some organo-species need first to be decomposed in order to achieve on-line volatile species generation (such on-line digestion is particularly interesting in metal speciation, for species decomposition at the interface between the exit of a liquid chromatographic column and the atomic detector). For these purposes, different principles such as on-line chemical oxidation at room temperature, photo-oxidation and microwave heating have been proposed.

7.7.1. On-line photo-oxidation flow systems

The use of UV radiation (in some cases in combination with a strong oxidizing agent like hydrogen peroxide or peroxodisulphate) to decompose organic matter is widely described in the literature. A mercury lamp is most commonly used as the UV source; while most UV photoreactors are made of quartz or PTFE. In addition to its low UV absorption, PTFE offers favourable properties (such as lower cost, easier handling and lower fragility) than quartz for constructing tubular photoreactors.

In the area of flow systems for analytical purposes, several photoreactor designs have been proposed in order to obtain appropriate efficiency [**Rubio**

***et al* (1995)**]. Usually, the lamp is wrapped with a coil of tubing (e.g., PTFE) through which the sample flows. To increase the light intensity reaching the coil and to prevent eye exposure to UV radiation, the unit is enclosed in aluminium foil.

7.7.2. On-line microwave-assisted digestion

A priori, microwave (MW) digestion in a flow was expected to be associated with serious problems derived from the vigorous chemical conditions, elevated temperatures, high pressures, and long digestion times frequently required to obtain complete decomposition. Furthermore, an additional problem to be faced is the evacuation of the gases produced during the digestion step. Fortunately, interesting instrumental designs have been developed that attempt to solve these problems [**Burguera *et al* (2001)**].

The MW oven can be incorporated into flow manifolds, which may be off-line or on-line connected to the atomic detector. For example, in order to analyze samples requiring long digestion times, the flow can be interrupted for a period of time, while the sample is in the oven, resulting in a stopped-flow digestion system. In some designs, the oven is modified by placing an electric fan to vent hot air during operation and to help cooling the tube at the end of the digestion. Frequently, a cooling system is on-line connected after the digestor. Also, in some designs a backpressure regulator or a diffusion cell connected to a vacuum pump is located prior the spectrometer in order to remove the fumes produced during acid decomposition of organic materials.

On-line microwave systems have been already proposed for a variety of samples, such as botanical and animal tissues, urine, sediments, whole blood, waters, etc. The samples can be injected into the flow manifold as slurries. It is interesting to note studies on trace element speciation where a MW oven is on-line located between a column for high performance liquid chromatography (HPLC) and a cold vapour or hydride generation system. The oven decomposes on-line separated organo-species (e.g., of arsenic, mercury or selenium) thus allowing an efficient on-line generation of the volatile derivatives (see section 7.10.3) and final robust elemental detection by AAS.

7.8. FLOW INJECTION COUPLED TO ELECTROTHERMAL ATOMISATION

The intrinsic discontinuous nature of the ETAAS limits the interest of interfacing continuous flow manifolds to this detector. However, some flow approaches offer special attractiveness for their semi-on-line combination with ETAAS [**Vereda Alonso *et al* (2001)**]. Of particular practical interest are:

a) *Formation of volatile derivatives of the analyte and their preconcentration in the graphite tube.* Continuously heated graphite tubes, wherein the volatile analyte hydride enters a hot furnace (1800-2300°C) and is atomised during its transit time through the device, have had limited use. However, *in situ* trapping techniques, which couple hydride generation with graphite furnace analyte collection, are a growing area of interest because a clean, rapid separation/preconcentration of the analyte from the matrix is achieved in a simple way, and automated commercial instruments are available for such purposes.

In situ trapping of previously volatilised species and final ETAAS detection has been successfully applied to the determination of As, Hg, Ge, Bi, Pb, In, Sb, Se, Sn, Cd and Te in a great variety of matrices.

In such systems, the graphite furnace is used as both the hydride trapping medium and the atomisation cell. The hydride purged from the generator is trapped in the pre-heated furnace, usually at 300-600°C, until the evolution of hydride is completed. The trapped analyte is subsequently atomized at temperatures generally over 2200°C. This technique has been shown to enhance the sensitivity significantly and to eliminate effectively the possible influence of the hydride generation kinetics on the signal shape. The nature of the graphite tube is expected to affect greatly the efficiency of hydride adsorption. It has been shown that the coating of the graphite tube with Pd, Zr, Ag or Pd-Ir mixtures, improves the sensitivities and precision significantly.

b) *Separation and preconcentration by on-line column sorption and co-precipitation.* The compatibility of organic solvents with ETAAS has prompted a number of interesting applications based on the sorption of metal chelates on non-ionic sorbents and their later elution with organic solvents such as ethanol, methanol and acetonitrile. Similarly, co-precipitation systems have also been synchronously coupled to ETAAS for the determination of trace amounts of heavy metals. The analytes are co-precipitated with bulky agents, such as the iron(II)-hexamethylenedithiocarbamate complex, on the walls of a knotted reactor. The precipitate is later dissolved in a few microlitres of isobutyl methyl ketone, stored in a PTFE tube, and delivered finally into the graphite tube atomiser for AAS measurement.

7.9. CHROMATOGRAPHIC SEPARATIONS COUPLED ON-LINE TO ATOMIC ABSORPTION SPECTROMETRY

Metal speciation concerns the identification and quantification of specific forms of a given metal. As different forms of an element may exhibit different toxicities and mobilities in the environment, it is of clear importance to be able to distinguish between the individual species present in a given sample [**Sanz-Medel (1998)**]. In this context, total determinations of a toxic element by atomic spectroscopy are insufficient today and sometimes misleading when assessing a biological impact, e.g. arsenobetaine is not toxic, methylmercury is much more toxic than inorganic mercury, and tributyltin is a most potent biocide, while Sn(IV) is not. Therefore, additional "speciation" information to complement a total toxic element determination is being increasingly demanded in nutritional, toxicological, environmental and clinical/biological fields.

Thus, analytical approaches able to provide reliable molecular information of the sought particular species of the element under study are needed for speciation purposes. Of course, atomic methods are by definition non-speciating methods as the atomiser destroys the molecules. However, the so-called "hybrid" techniques, consisting of the coupling of a powerful separation (generally chromatographic) technique with a sensitive element-specific atomic detector offer the ideal combination. For discontinuous atomic detectors such as ETAAS, the separation can be carried out "off-line" with the detector. However, it is more convenient to on-line couple the separation unit to a continuous detector, such as FAAS, ICP-OES or ICP-MS.

The selection of the separation technique depends on the nature and physical properties of the species of the element to be determined. When volatile, thermostable, neutral species (or the ability to produce them by chemical derivatization) have to be determined, the technique of choice is gas chromatography. For example, Grignard reagents such as butylmagnesium chloride dissolved in tetrahydrofuran are used to convert methyl- and inorganic mercury to volatile non-polar butylmercury derivatives. Alternative derivatization reagents of great interest are sodium tetraalkylborates (such as sodium tetraethylborate) because they allow one to carry out the derivatization process in an aqueous medium.

For non-volatile, thermally unstable, or charged compounds, the selection can be by the more versatile but less sensitive HPLC. Provided that the target metal(loid) forms volatile derivatives, the exit of the column is on-line connected to a gas-liquid flow generator system. For a species of metal(loid) that do not form a volatile derivative, approaches to on-line decomposition of eluted species using oxidizing agents, UV radiation or MW energy prior to the gas generation step, have been successfully tested (see section 7.10.3), before the AAS measurements.

Finally, just a few words here to note that some promising work in chromatographic detection by AAS is being carried out using a diode laser as a light source and flames or plasmas as atomizers [**Zybin *et al* (2004)**]. Many HPLC methods employ high concentrations of organic solvents or salts and an important advantage of FAAS is that the flame is more robust against these compounds compared to a plasma (e.g., in ICP-MS), thereby providing higher flexibility for appropriate chromatographic conditions selection. The use of diode lasers for FAAS typically improves the sensitivity of the technique by one to three orders of magnitude while maintaining the advantages of low cost, robustness and easy of use, thus providing an alternative to complicated and expensive detection systems for speciation analysis. Elements such as Cl (837.82 nm), C (833.74 nm) and H (656.45 nm) can be easily monitored by AAS using diode lasers. In this context, diode laser AAS in combination with plasmas, such as the microwave induced plasma (MIP), seems to offer a high potential as a gas chromatographic detector of organohalides.

7.10. APPLICATIONS AND CASE STUDIES

The three case studies following illustrate possible applications of the combination of flow systems to AAS.

7.10.1. On-line aluminium pre-concentration and its application to the determination of the metal in dialysis concentrates

Certain clinical disorders that have been identified in renal-failure patients undergoing regular dialysis are associated with aluminium loading in the human body. For this reason, the solutions used for dialysis treatment should be checked for very low levels of Al. However, the determination of Al in concentrates for haemodialysis constitutes a formidable challenge since a low concentration of the metal must be determined in a dramatically high concentration of inorganic salts (e.g. about 10^5 mg.L^{-1} of sodium ions and 10^3 mg.L^{-1} of calcium ions).

This problem was successfully overcome some time ago by the use of an on-line minicolumn packed with Chelex® 100. Figure 7.10 shows the flow manifold. The set-up consisted of a peristaltic pump (A), a septum for sample injection (B), a mixing coil, a second valve for eluent injection (C), a minicolumn (D) and a third valve (E) placed after the minicolumn to minimise clogging of the nebuliser. As shown in Figure 7.10a, the sample injected through the septum mixes with the carrier along the mixing coil and the aluminium is retained on the minicolumn. During this part of the cycle the non-retained matrix salts go to waste and water is continuously pumped to the detector. Figure 7.10b shows the step that follows once the preconcentration cycle is completed. First, valve E is turned and then the releasing solution (100 μL of 2M HCl) is inserted into the flow stream by turning valve C, hence eluting Al directly into the nebuliser of the spectrometer [**Pereiro *et al* (1990)**].

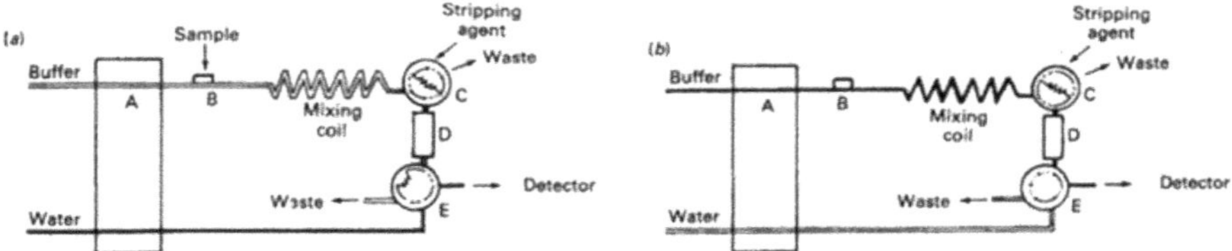

Figure 7.10. Flow diagram of the system used for the determination of Al. (a) Pre-concentration step; and (b) elution step. A, peristaltic pump; B, C and E, injection valves; D, minicolumn. In each figure the black flow line corresponds to the channel entering the nebuliser (from M.R. Pereiro, A. Lopez, M.E. Diaz, A. Sanz-Medel, J. Anal. At. Spectrom., 5 (1990) 15. Reproduced with permission of the Royal Society of Chemistry).

This configuration not only prevents clogging of the nebuliser, but also allows careful control of optimum experimental conditions for the determination of Al (during the preconcentration step, a standard solution of the metal can by pumped instead of water through the secondary line for detector control purposes). Using this manifold, a detection limit of 15 $\mu g.L^{-1}$ was obtained with FAAS detection for 1 mL sample injections. Of course, this detection limit can be improved by preconcentrating higher volumes of sample. The methodology was successfully tested for the determination of Al in dialysis concentrates.

7.10.2 Indirect atomic absorption spectrometric determination of sulphonamides in pharmaceutical preparations and urine

The use of flow separation systems allows important increases of the analytical reproducibility achievable by atomic spectrometry indirect determinations. The indirect determination of many compounds such as cocaine, methadone, bromazepan, salicylic acid, cloramphenicol, etc, have been successfully achieved with flow manifolds on-line coupled to atomic absorption spectrometers [**Yebra (2000)**].

Figure 7.11 shows a diagram of the manifold proposed for the determination of sulphonamides in pharmaceutical preparations and urine. Cu(II) was injected into a carrier solution. Firstly, the carrier solution consisted of water, and a signal corresponding to copper was obtained. In the next step, the carrier was the sample solution containing sulphonamide and the Cu(II) solution was again injected. The reaction of this cation with the analyte gave rise to a precipitate, which is on-line retained on a filter, and a smaller positive peak was then obtained because of the precipitate formation. The difference between the two peaks corresponds to the amount of precipitated copper, which can be related to the sulphonamide concentration in the sample. A linear range for the determination of sulphamethazine between 2.5 and 35 $mg.L^{-1}$ was obtained. The method has been successfully applied to the determination of total sulphonamides in eight different pharmaceutical preparations and to the determination of sulphamethazine in urine.

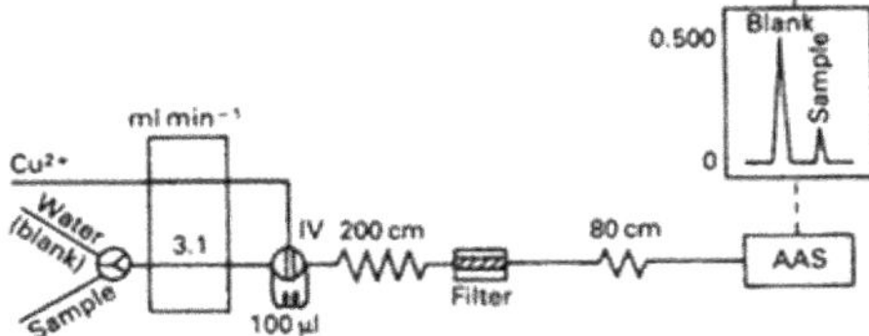

Figure 7.11. Manifold and optimum working conditions used in the determination of sulphonamide. IV, injection valve (from R. Montero, M. Gallego, M. Valcarcel, J. Anal. At. Spectrom., 3 (1988) 725. Reproduced with permission of the Royal Society of Chemistry).

7.10.3. Coupled high performance liquid chromatography–microwave digestion–hydride generation–atomic absorption spectrometry (HPLC-MW-HG-AAS) for inorganic and organic arsenic speciation in fish tissue

Arsenic is present in the environment in a number of different inorganic and organic forms, due to its participation in complex biological and chemical processes. As it was described in Chapter 5 (section 5.5.1), naturally occurring arsenic species in water and soil include: As(III), As(V), monomethylarsonic acid (MMA) and dimethylarsinic acid (DMA). Other organoarsenic compounds such as arsenobetaine (AsB), arsenocholine (AsC) and arsenosugars may occur in biological tissues.

Different arsenic species show different toxicity and chemical behaviour, leading to differences in their metabolism. Although complex organoarsenic compounds such as AsB and AsC are considered to be tolerated by living organisms, it has been reported that under specific conditions AsB present in marine biomass could be transformed into trimethylarsine, which is a toxic species, and therefore the determination of AsB and AsC in foods is important.

Some of the most suitable methods proposed in the literature to carry out arsenic speciation are based on the combination of HPLC with on-line detection by hydride generation coupled to an atomic spectrometer. The introduction of the hydride generation as a post-column derivatization step leads to increased sensitivity and matrix removal, because the volatile hydrides are separated from the liquid residue. However, when hydride generation is employed for arsenic speciation it must be borne in mind that some bioorganoarsenic compounds, such as AsB, AsC and arsenosugars, are not detected by this approach, because they are resistant to acid digestion and do not form volatile hydrides. However, the use of on-line microwave digestion allows for fast decomposition of the species with great operating simplicity. Figure 7.12 shows an HPLC-MW-HG-AAS system for arsenic speciation in fish tissue ($K_2S_2O_8$ is used as the oxidant).

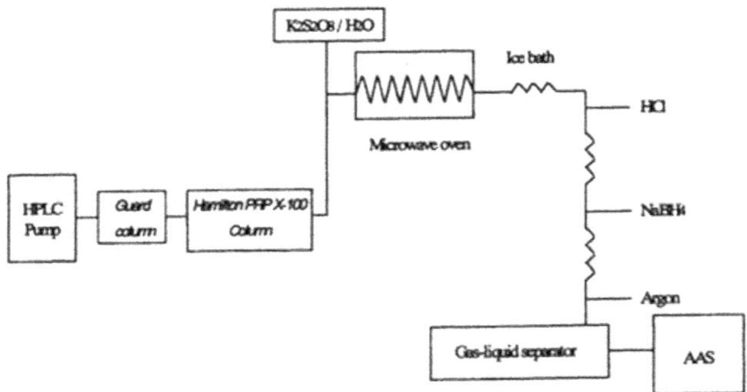

Figure 7.12. HPLC-MW-HG-AAS system for arsenic speciation (from M.C. Villa-Lojo, E. Alonso-Rodriguez, P. Lopez-Mahia, S. Muniategui-Lorenzo, D. Prada-Rodriguez, Talanta, 57 (2002) 741. Reproduced with permission of Elsevier).

It was observed than when HPLC is used for arsenic species separation, AsB and As(III) peaks overlap under the chromatographic conditions used with most columns. To solve this problem, two separate injections, one using an oxidation step and the other without it, were proposed. Other alternatives consist of the previous isolation of AsB and AsC from the other arsenic species (As(III), As(V), MMA and DMA) using a Sep-Pak® solid-phase extraction (SPE) cartridge. The retained AsB and AsC are later eluted and injected in the chromatographic system.

Chapter Eight

BUYERS GUIDE AND REFERENCE SECTION

8.1. COMPANIES

A list of ten manufacturers (alphabetic order) of atomic absorption spectrometers serving worldwide is provided below:

Analytik Jena AG

Konrad-Zuse-Straβe 1

07745 Jena, Germany

Tel: (49) 3641 77 70

e-mail: info@analytik-jena.de Web: www.analytik-jena.de

Aurora Instruments Ltd

1001 East Pender Street

Vancouver, B.C., V6A 1W2 Canada

Tel: (1) 604 215 8700

e-mail: info@aurora-instr.com Web: www.aurora-instr.com

Buck Scientific, Inc.

Headquarters

58 Fort Point Street.

East Norwalk, CT06855, USA

Tel: (1) 203 853 9444

e-mail: sales@bucksci.com Web: www.bucksci.com

GBC Scientific Equipment Pty Ltd.

Headquarters

Monterey Rd

Dandenong, VIC (Australia)

Tel: (61) 3 9213 3666

e-mail: gbc@gbcsci.com Web: www.gbcsci.com

Hitachi, Ltd.

Principal Office

6-6, Marunouchi 1-chome, Chiyoda-ku

Tokio, 100-8280 Japan.

Phone : (81) 3 3258 1111 Web : www.hitachi.com

Teledyne Leeman Labs, Inc.

6 Wenthworth Drive

Hudson, NH 03051

Tel: (1) 603 886 8400

e-mail: SalesInfo@LeemanLabs.com Web: www.leemanlabs.com

Perkin Elmer, Inc.

Corporate Headquarters

45 William Street

Wellesley, MA02481-4078, USA

Tel: (1) 781 237 5100

e-mail: ProductInfo@perkinelmer.com Web: www.perkinelmer.com

Shimadzu Corporation

Head Office

1, Nishinokyo-Kuwabara-cho, Nakagyo-ku

Kyoto 604-8511, Japan

Tel: (81) 75 823 1111 Web: www.shimadzu.com

Thermo Electron Corporation

Corporate Headquarters

81 Wyman Street

Waltham, MA 02454, USA

Tel: (1) 781 622 1000 Web: www.thermo.com

Varian, Inc.

Corporate Headquarters

3120 Hansen Way, Palo Alto, CA 94304-1030, USA

Tel: (1) 650 213 8000

e-mail: customercare@varianinc.com Web: www.varianinc.com

8.2. GLOSSARY OF TERMS

[Terms in *italics* are further defined within the Glossary.]

Absorbance: the logarithmic function of the inverse of *transmittance.*

Absorptivity: also known as the molar extinction coefficient in molecular spectroscopy, it is the wavelength-dependent absorption of an analyte as a function of concentration and path length. It is expressed in units of $concentration^{-1} \cdot cm^{-1}$.

Analyte: a sample component whose concentration needs to be known.

Ashing: or pyrolysis, is an step in an ETAAS program that is designed to remove matrix constituents that might interfere with the measurement of the analyte.

Atomization: the process of producing gaseous atoms for the atomic measurement. The atom-forming process usually requires a high temperature (except for cold-vapor methods), which is usually produced by a flame or by electrical current flowing through a resistive medium.

Atomization efficiency: Efficiency to produce atomic vapor.

Background absorption: absorption of the source radiation by the flame itself and/or by concomitant species introduced into the atomizer.

Bandpass: see *Spectral bandwidth.*

Bandwidth: see *Spectral bandwidth.*

Blank: an "ideal blank" contains all the sample constituents except the analyte. However, it is difficult to prepare an ideal blank because the concomitants and their concentrations are not usually known. In some cases it is sufficient to use just the same solvent and any reagents employed to condition the sample.

Blank signal: includes the background signal due to optical signals from the sample container and the concomitants in the blank.

Beer-Lambert law: the law that defines a linear relationship between concentration and absorbance. It is often written as "absorbance = absorptivity x path-length x concentration".

Calibration: a quantitative procedure performed to relate the known concentration of standards containing the analyte to the detector signal.

Calibration curve: also known as "working curve" and corresponds to the relationship of instrument response (absorbance) as a function of concentration.

Cold-vapor: the chemical method by which atoms of mercury or cadmium are produced from a solution containing Hg or Cd ions.

Detector: the part of the instrument that converts radiant energy from the light source to electricity. A *photomultiplier tube* is typically used, but it may also be a solid-state detector in modern instrumentation.

Detection limit: is indicative of the lowest amount of an analyte that can be detected with a specified degree of certainty. This is most often defined as three times the standard deviation of the blank measurement.

Deuterium background corrector: a strategy to correct for background absorption. This method employs a continuum radiation source (the deuterium lamp) that passes through the atomizer.

Diffraction grating: a plane or concave plate that is ruled with closely spaced grooves. The grating acts like a multislit source when collimated radiation strikes it. Different wavelengths are diffracted at different angles.

Double-beam optics: the optical design whereby a percentage of the radiation from the light source of an atomic absorption spectrometer is diverted before it reaches the atomization cell and monitored to compensate for drift in light source intensity.

Drain trap: A hole at the bottom of the mixing chamber that leads through a plastic tube to a water-filled trap, allowing waste sample solution to drain from the mixing chamber but not allowing combustion gases to escape.

Dynamic range: sometimes known as "linear dynamic range" or "linear range", the analyte concentration range over which response is a well defined (usually linear) function of the analyte concentration.

Electrodeless discharge lamp: an intense radiation source for AAS, which consists of a sealed quartz tube containing a small amount of the element of interest and an inert gas. The lamp is placed in a radiofrequency field, which excites the atoms to emit intense line radiation. These sources are usually used for elements such as As, Se, Hg and Sb.

Flow-rate (solution): the volumetric flow-rate ($mL.min^{-1}$) of solution uptake into the nebulizer of a FAAS instrument.

Flow-rate (gas): the volumetric flow-rate of combustible gases ($L.min^{-1}$) into the mixing chamber of a flame AAS instrument, or of inert gas used for graphite furnace and gas generation methods.

Flow spoiler: a plastic, fan-shaped device placed in the mixing chamber of a flame AAS to improve the mixing of combustion gases and analyte solution droplets and facilitate the removal of large droplets down the drain trap at the bottom of the mixing chamber.

Focal plane: a plane onto which the spectrum is dispersed. The plane contains an aperture(s) or slit(s), with the photodetector(s) located on the other side of the aperture(s) or slit(s).

Graphite tube: the most common atomization cell used in an electrothermal atomizer for AA. Typically made of *pyrolytic graphite* and bathed in an inert gas such as argon to prevent decomposition. Can be heated up to 3000°C.

Ground state: the lowest energy state of an atom or molecule.

Half-width: also called "full width at half maximum" (FWHM), is the width in wavelength units at half the net peak height.

Hollow cathode lamp: the most common radiation source for atomic absorption. It consists of a low-pressure inert-gas-filled tube containing an anode and a hollow cathode made from the element for which the lamp is to produce atomic line radiation.

Hydride-generation: the method by which hydride-forming elements, such as As, Se, Sb, and Te are released from solutions of their ions, commonly using sodium borohydride. The released hydrides are swept from solution by an inert gas and decomposed to atoms in an absorption cell at a temperature of approximately 1000°C.

Integration: a process for identifying and calculating the amount of a component by measuring the area greater than the baseline, defined by the instrument blank, over a specific time period. In conventional FAAS, integration times of three to ten seconds are most commonly used, since continuous signals are measured. In ETAAS, flow injection, and gas-generation methods that produce transient signals, the "*peak*" produced by the analyte is integrated above the baseline.

Interferent: is a substance present in the analytical sample which affects the magnitude of spectral signal measured for the analyte.

Linear dispersion: informs how far apart in distance are separated two close wavelengths in the focal plane of a dispersive wavelength selector. Linear dispersion is conveniently expressed in $mm.nm^{-1}$.

Matrix: components present in the sample that are not the analyte(s).

Matrix modifier: an element or compound that is added to the sample in ETAAS in order to increase the volatility of the matrix (and thus remove it during the ashing stage of the temperature program), or to decrease the volatility of the analyte element so that it can be atomized at high temperatures.

Memory effects: influence exerted by the previously measured sample (or standard) on the signal generated by the current sample (or standard) being measured.

Metal speciation: identification and quantitation of specific chemical forms of a given metal.

Mixing chamber: a chamber in which combustible gases are mixed with the solution droplets from the nebulizer and then transported to the flame. Larger droplets (approximately 95% of the sample) are removed from the mixing chamber through the *drain trap*.

Modulation: the periodic variation of the radiation from the light source, introduced either electronically or mechanically with a chopper. Modulation of the light source allows the instrument to discriminate against other sources of radiation that might reach the detector and bias the absorbance measurement.

Monochromator: a wavelength selection device used in atomic absorption spectrometers to isolate the absorbable radiation (analytical wavelength or wavelength region) from other radiation.

Nebulizer: component of a flame atomic absorption sample introduction system that draws aqueous solution into the premix chamber and converts it to a fine mist of small droplets that are swept into the flame.

Noise: the variation in the signal produced by the instrument. Noise is caused by short and long-term variations in different instrument components.

Parts per billion or ppb: in a liquid corresponds to micrograms of analyte per liter of sample ($\mu g.L^{-1}$) and in a solid to nanograms of analyte per gram of sample ($ng.g^{-1}$).

Parts per million or ppm: in a liquid corresponds to milligrams of analyte per liter of sample ($mg.L^{-1}$) and in a solid to micrograms of analyte per gram of sample ($\mu g.g^{-1}$).

Peak: the transient increase in atomic absorption, using flow injection procedures, gas-generation methods or electrothermal atomizers, the height and area of which are related to the concentration of analyte element in a sample.

Photomultiplier tube: the most often used detector in an atomic absorption spectrometer. It consists of a vacuum tube containing an alkali-element photocathode that produces electrons when struck by photons of sufficient energy (the photoelectric effect). Each photoelectron is then multiplied by collisions with a series of dynodes so that the electrical signal produced by each photon is greatly amplified.

Platform atomization: also known as the L'vov platform, is a small platform onto which a sample is placed inside the graphite furnace tube rather than placing the sample on the tube wall. This delays sample atomization until the gas temperature inside the furnace is higher than it would be for wall atomization, which reduces some interference effects.

Pyrolytic graphite: a form of graphite manufactured by decomposition of methane gas at low pressure (about 1 torr) and heated to 2000°C. The result is a very highly ordered, chemically inert, impermeable, highly pure and stable to 3000°C product.

Qualitative analysis: the identification of the components in the sample.

Quantitative analysis: the determination of the amount or concentration of one/several component(s) of the sample.

Refractory elements in AAS: Elements that tend to form heat-stable compounds in the flame.

Resonance transition: a transition to or from the ground electronic level.

Resonance line: the resulting spectral line from a *resonance transition*.

Reciprocal linear dispersion: represents the number of wavelength intervals (e.g. nm) contained in each interval of distance (e.g. mm) along the focal plane.

Sample throughput: is the number of samples that can be analyzed, or elements that can be determined, per unit time.

Smith-Hieftje background corrector: a method to correct for background absorption in AAS that pulses the hollow cathode lamp at low and then at high current. During the high current pulse, a large cloud of atoms is formed in front of the lamp cathode. This cloud essentially prevents absorbable radiation from reaching the analyte in the atomizer and thus allows discrimination of atomic absorption from other sources of absorption.

Spectral bandwidth: or spectral bandpass, is the difference between wavelengths at which the intensity is half the peak intensity passed by the exit slit of a dispersive wavelength selector.

Spectrometry: quantitative measurement of the intensity of electromagnetic radiation at one or more wavelengths with a photoelectric detector.

Spectrum: in optical spectroscopy corresponds to the display of the intensity of radiation emitted, absorbed or scattered by a sample *versus* a quantity related to photon energy, such as wavelength or frequency.

Sputtering: a process in which atoms or ions are ejected from a surface by a beam of charged particles.

Standard: solution or solid with a known concentration of analyte (used for instrument calibration).

Standard addition: a calibration method that compensates for matrix-induced enhancement or suppression of analyte signals. A known concentration of analyte element is added to the sample and the instrument response of the known concentration of added element is used to calibrate the instrument response for the sample.

Sensitivity: slope of the linear plot of "instrument response *versus* analyte concentration". Traditionally, in AAS, the sensitivity is defined as the concentration of analyte that produces an instrument response of 0.0044 absorbance units (1% absorption).

Transmittance: the fraction of the incident electromagnetic radiation that is transmitted by a sample.

Validated method: a method that meets or exceeds certain sampling and measurement performance criteria.

Zeeman background corrector: a method to correct for background absorption in atomic absorption that uses a magnetic field.

8.3. STANDARDS

8.3.1. British Standards Institution

In many circumstances British Standards (BS) standards are automatically incorporated from standards promoted by the International Standards Organization (ISO). The European Committee for Standardization (CEN) and the European Committee for Electrotechnical Standardization (CENELEC) elaborates or adopts technical standards to promote free trading, the safety of workers and consumers, interoperability of networks, environmental protection, exploitation of research and development programs, and public procurement. EN corresponds to the acronym of "European Norms". Whenever a BS norm is a transposition of an EN, this should be indicated as follows:

BS EN 1811:1999 – Reference test method for release of nickel from products intended to come into direct and prolonged contact with the skin [12 pp].

In the case that the adopted EN norm proceeds from and ISO norm, this should be indicated as follows:

BS EN ISO 6869:2001 – Animal feeding stuffs. Determination of the contents of calcium, copper, iron, magnesium, manganese, potassium, sodium and zinc. Method using atomic absorption spectrometry [26 pp].

Below are collected a few examples of BS standards not originating from EN or ISO standards.

BS 2000-455:2000 – Methods of test for petroleum and its products. Determination of manganese in gasoline. Atomic absorption spectrometry (AAS) method [4 pp].

BS 5766-6:1984 – Methods for analysis of animal feeding stuffs. Determination of calcium by atomic absorption spectrometry [8 pp].

BS 6075-11:1981 - Methods of sampling and test for sodium hydroxide for industrial use. Determination of mercury content (flameless atomic absorption method) [10 pp].

BS 7164-28.1:1977 - Chemical tests for raw and vulcanized rubber. Methods for determination of copper content. Atomic absorption spectrometry [14 pp].

BS 7317-3:1990 - Methods for analysis of high purity copper Cu-CATH-1. Method for determination of antimony, arsenic, bismuth, selenium, tellurium and tin by hydride generation and atomic absorption spectrometry [12 pp].

8.3.2. International Standards Organization

There is a wide variety of ISO Standards on atomic absorption spectrometry for analysis of elements in food, environment, raw materials, industrial products, etc. Below a small selection of methods is collected together that uses either FAAS, ETAAS, generation of hydride-AAS or cold vapor-AAS:

ISO 4744:1984 – Copper and copper alloys; Determination of chromium content; Flame atomic absorption spectrometric method [3 pp].

ISO 5373:1981 – Condensed phosphates for industrial use (including foodstuffs); Determination of calcium content; Flame atomic absorption spectrometric method [4 pp].

ISO 5889:1983 – Manganese ores and concentrates; Determination of aluminium, copper, lead and zinc contents; Flame atomic absorption spectrometric method [4 pp].

ISO 6636-2:1981 – Fruit, vegetables and derived products – Determination of zinc content – Part 2: atomic absorption spectrometric method [4 pp].

ISO 7530-1:1990 – Nickel alloys; flame atomic absorption spectrometric analysis; part 1: general requirements and sample dissolution [7 pp].

ISO 7530-8:1992 – Nickel alloys; flame atomic atomic absorption spectrometric analysis; part 8: determination of silicon content [3 pp].

ISO 7627-4:1983 – Hardmetals; Chemical analysis by flame atomic atomic absorption spectrometry; Part 4: Determination of molybdenum, titanium and vanadium contents from 0.01 to 0.5 % (m/m) [2 pp].

ISO 8294:1994 – Animal and vegetable fats and oils – Determination of copper, iron and nickel contents – Graphite furnace atomic absorption method [6 pp].

ISO 8658:1997 – Carbonaceous materials for use in the production of aluminium – Green and calcined coke – Determination of trace elements by flame atomic absorption spectroscopy [8 pp].

ISO 9174:1998 – Water quality – Determination of chromium – Atomic absorption spectrometric methods [10 pp].

ISO 9668:1990 – Pulps; determination of magnesium content; flame atomic absorption spectrometric method [3 pp].

ISO 9965:1993 – Water quality; determination of selenium; atomic absorption spectrometric method (hydride technique) [5 pp].

ISO 10136-3:1993 – Glass and glassware; analysis of extract solutions; part 3: determination of calcium oxide and magnesium oxide by flame atomic absorption spectrometry [6 pp].

ISO 11047:1998 - Soil quality - Determination of cadmium, chromium, cobalt, copper, lead, manganese, nickel and zinc in aqua regia extracts of soil – Flame and electrothermal atomic absorption spectrometric methods [18 pp].

ISO 11438-1:1993 – Ferronickel; determination of trace-element content by electrothermal atomic absorption spectrometric method; part 1: general requirements and sample dissolution [5 pp].

ISO 11438-8:1993 – Ferronickel; determination of trace-element content by electrothermal atomic absorption spectrometric method; part 8: determination of indium content [3 pp].

ISO 11813:1998 – Milk and milk products – Determination of zinc content – Flame atomic absorption spectrometric method [6 pp].

ISO 12193:2004 – Animal and vegetable fats and oils – Determination of lead by direct graphite furnace atomic absorption spectroscopy [7 pp].

ISO 12740:1998 – Lead sulfide concentrates – Determination of silver and gold contents – Fire assay and flame atomic absorption spectrometric method using scorification and cupellation [26 pp].

ISO 14377:2002 – Canned evaporated milk – Determination of tin content – Method using graphite furnace atomic absorption spectrometry [8 pp].

ISO 15248:1998 – Zinc sulfide concentrates – Determination of silver and gold contents – fire assay and flame atomic absorption spectrometric method using scorification or cupellation [25 pp].

ISO 15774:2000 – Animal and vegetable fats and oils – Determination of cadmium content by direct graphite furnace atomic absorption spectrometry [6 pp].

ISO 17191:2004 – Urine-absorbing aids for incontinence – Measurement of airbone respirable polyacrylate superabsorbent materials – Determination of dust in collection cassettes by sodium atomic absorption spectrometry [9 pp].

ISO 17239:2004 – Fruits, vegetables and derived products – Determination of arsenic content – Method using hydride generation atomic absorption spectrometry [10 pp].

ISO 17733:2004 – Workplace air – Determination of mercury and inorganic mercury compounds – Method by cold vapor atomic absorption spectrometry or atomic fluorescence spectrometry [51 pp].

8.4. REFERENCES

Alexiu, V., Vladescu, L. (2005) – Optimization of a chemical modifier in the determination of selenium by graphite furnace atomic absorption spectrometry and its application to wheat and wheat flour analysis, *Analytical Sciences,* 21, 2005, pp. 137-141.

Arbab-Zavar, M.H., Chamsaz, M., Youssefi, A., Aliakbari, M. (2006) – Mechanistic aspects of electrochemical hydride generation for cadmium, *Analytica Chimica Acta,* 576, 2006, pp. 215-220.

Astrom, O. (1982) - Flow injection analysis for the determination of bismuth by atomic absorption spectrometry with hydride generation, *Analytical Chemistry,* 54, 1982, pp. 190-193.

Barbosa, F., Lima, E.C., Krug, F.J. (2000) – Determination of arsenic in sediment and soil slurries by electrothermal atomic absorption spectrometry using W-Rh permanent modifier, *Analyst,* 125, 2000, pp. 2079-2083.

Becker-Ross, H., Okruss, M., Florek, S., Heitmann, U., Huang, M.D. (2002) - Echelle-spectrograph as a tool for studies of structured background in flame atomic absorption spectrometry, *Spectrochimica Acta Part B,* 57, 2002, pp. 1493-1504.

Bolea, E., Arroyo, D., Laborda, F., Castillo, J.R. (2006) – Determination of antimony by electrochemical hydride generation atomic absorption spectrometry in samples with high iron content using chelating resins as on-line removal system, *Analytica Chimica Acta,* 569, 2006, pp. 227-233.

Bolshakov, A.A., Ganeev, A.A., Nemets, V.M. (2006) – Prospects in analytical atomic spectrometry, *Russian Chemical Reviews,* 75, 2006, pp. 289-302.

Burguera, J.L., Burguera, M. (2001) - Flow injection-electrothermal atomic absorption spectrometry configurations: recent developments and trends, *Spectrochimica Acta Part B,* 56, 2001, pp. 1801-1829.

Burguera, J.L., Burguera, M. (2004) – Analytical applications of organized assemblies for on-line spectrometric determinations: present and future, *Talanta,* 64, 2004, pp. 1099-1108.

Cal-Prieto, M.J., Felipe-Sotelo, M., Carlosena, A., Andrade, J.M., López-Mahía, P., Muniategui, S., Prada, D. (2002) - Slurry sampling for direct analysis of solid materials by electrothermal atomic absorption spectrometry. A literature review from 1990 to 2000, *Talanta,* 56, 2002, pp. 1-51.

Carmel, V. (2003) - Solid phase extraction of trace elements. Review, *Spectrochimica Acta Part B,* 58, 2003, pp. 1177-1233.

Chan, C.C.Y. (1985) – Semiautomated method for determination of selenium in geological materials using a flow injection analysis technique, *Analytical Chemistry*, 57, 1985, pp. 1482-1485.

D'Ulivo, A. (2004) – Chemical vapor generation by tetrahydroborate(III) and other borane complexes in aqueous media. A critical discussion of fundamental processes and mechanisms involved in reagent decomposition and hydride formation, *Spectrochimica Acta Part B*, 59, 2004, pp. 793-825.

Economou, A. (2005) - Sequential-injection analysis (SIA): A useful tool for on-line sample handling and pre-treatment, *Trends in Analytical Chemistry*, 24, 2005, pp. 416-425.

Emig, M., Billmers, R.I., Owens, K.G., Cernansky, N.P., Miller, D.L., Narducci, F.A. (2002) – Sensitive and selective detection of paramagnetic species using cavity enhanced magneto-optic rotation, *Applied Spectroscopy*, 56, 2002, pp. 863-868.

Fernandez de la Campa, M.R., Segovia Garcia, E., Valdes-Hevia y Temprano, M.C., Aizpun Fernandez, B., Marchante Gayon, J.M., Sanz-Medel, A. (1995) – Effects of organised media on the generation of volatile species for atomic spectrometry, *Spectrochimica Acta Part B*, 50, 1995, pp. 377-391.

Frech, W. (1996) – Recent developments in atomizers for electrothermal atomic absorption spectrometry, *Fresenius Journal of Analytical Chemistry*, 355, 1996, pp. 475-486.

Grinberg, P., Calixto de Campos, R. (2001) – Iridium as permanent modifier in the determination of lead in whole blood and urine by electrothermal atomic absorption spectrometry, *Spectrochimica Acta Part B*, 56, 2001, pp. 1831-1843.

Guo, X.W., Guo X.M. (1995) – Determination of cadmium at ultratrace levels by cold vapour atomic absorption spectrometry, *Journal of Analytical Atomic Spectrometry*, 10, 1995, pp. 987-991.

Hanna, C.P., Haigh, P.E., Tyson, J.F., McIntosh, S. (1993) – Examination of separation efficiencies of mercury vapour for different gas-liquid separators in flow injection cold vapour atomic absorption spectrometry with amalgam preconcentration, *Journal of Analytical Atomic Spectrometry*, 8, 1993, pp. 585-590.

Heitmann, U., Schütz, M., Becker-Ross, H., Florek, S. (1996) – Measurements of the Zeeman-spliting of analytical lines by means of a continuum source graphite furnace atomic absorption spectrometer with a linear charge coupled device array, *Spectrochimica Acta Part B,* 51, 1996, pp. 1095-1105.

Hieftje, G.M. (1989) – Atomic absorption spectrometry – Has it gone or where is it going?, *Journal of Analytical Atomic Spectrometry,* 4, 1989, pp. 117-122.

Hywel Evans, E., Day J.A., Palmer, C.D., Price, W.J., Smith, C.M.M., Tyson, J.F. (2005) – Atomic spectrometry update. Advances in atomic emission, absorption and fluorescence spectrometry, and related techniques, *Journal of Analytical. Atomic Spectrometry,* 20, 2005, pp. 562-590.

Hywel Evans, E., Day J.A., Palmer, C., Price, W.J., Smith, C.M.M., Tyson, J.F. (2006) – Atomic spectrometry update. Advances in atomic emission, absorption and fluorescence spectrometry, and related techniques, *Journal of Analytical Atomic Spectrometry,* 21, 2006, pp. 592-625.

Inczédy J., Lengyel T., Ure A.M. (1998) - *International Union of Pure and Applied Chemistry. Compendium of Analytical Nomenclature. Definitive rules 1997, 3rd Edition,* Blackwell Science, Oxford, England, 1998, ISBN: 0-86542-615-5, 964 pp.

Lenehan, C.E., Barnett, N.W., Lewis, S.W. (2002) - Sequential injection analysis, *Analyst,* 127, 2002, pp. 997-1020.

Montero, R., Gallego, M., Valcárcel, M. (1988) – Indirect atomic absorption spectrometric determination of sulphonamides in pharmaceutical preparations and urine by continuous precipitation, *Journal of Analytical Atomic Spectrometry,* 3, 1988, pp. 725-729.

Nóbrega, J.A., Rust, J., Calloway, C.P., Jones, B.T. (2004) - Use of modifiers with metal atomizers in electrothermal atomic absorption spectrometry: a short review, *Spectrochimica Acta Part B,* 59, 2004, pp. 1337-1345.

Ortner, H.M., Bulska, E., Rohr, U., Schlemmer, G., Weinbrunch, S., Welz, B. (2002) - Modifiers and coatings in graphite furnace atomic absorption spectrometry – mechanisms of action (a tutorial review), *Spectrochimica Acta Part B,* 57, 2002, pp. 1835-1853.

Pereiro, M.R., López, A., Díaz, M.E., Sanz-Medel, A. (1990) – On-line aluminium preconcentration and its application to the determination of the metal in dialysis concentrates by atomic spectrometric methods, *Journal of Analytical Atomic Spectrometry,* 5, 1990, pp. 15-19.

Parsons, P.J., Slavin, W. (1993) – A rapid Zeeman graphite furnace atomic absorption spectrometric method for the determination of lead in blood, *Spectrochimica Acta Part B,* 48, 1993, pp. 925-939.

Rapsomanikis, S. (1994) - Derivatization by ethylation with sodium tetraethylborate for the speciation of metals and organometallics in environmental samples. A review, *Analyst,* 119, 1994, pp. 1429-1439.

Rubio, R., Albertí, J., Padró, J., Rauret, G. (1995) – On-line photolytic decomposition for the determination of organoarsenic compounds, *Trends in Analytical Chemistry,* 16, 1995, pp. 274-279.

Ruzicka, J., Marshall, G.D. (1990) – Sequential injection: a new concept for chemical sensors, process analysis and laboratory Assays, *Analytica Chimica Acta,* 237, 1990, pp. 329-343.

Sanz-Medel, A., Valdés-Hevia y Temprano, M.C., Bordel García, N., Fernández de la Campa, M.R. (1995) - Generation of cadmium atoms at room temperature using vesicles and its application to cadmium determination by cold vapour atomic spectrometry, *Analytical Chemistry,* 67, 1995, pp. 2216-2223.

Sanz-Medel, A. (1998) - Toxic trace metal speciation: importance and tools for environmental and biological analysis, *Pure & Applied Chemistry,* 70, 1998, pp. 2281-2285.

Síma, J., Rychlovsky, P., Dedina, J. (2004) – The efficiency of the electrochemical generation of volatile hydrides studied by radiometry and atomic absorption spectrometry, *Spectrochimica Acta Part B,* 59, 2004, pp. 125-133.

Slavin, W., Manning, D.C., Carnrick, G.R. (1981) – The stabilized temperature platform furnace, *Atomic Spectroscopy,* 2, 1981, pp. 137-145.

Stafilov, T. (2000) - Determination of trace elements in minerals by electrothermal atomic absorption spectrometry, *Spectrochimica Acta Part B,* 55, 2000, pp. 893-906.

Sturgeon, R.E. (1996) – The graphite furnace and its role in atomic spectroscopy, *Fresenius Journal of Analytical Chemistry,* 355, 1996, pp. 425-432.

Sturgeon, R.E., Mester, Z. (2002) – Analytical applications of volatile metal derivatives, *Applied Spectroscopy,* 56, 2002, pp. 202A-213A.

Theodoroela, S., Thomaidis, N.S., Piperaki, E. (2005) – Determination of selenium in human milk by electrothermal atomic absorption spectrometry and chemical modification, *Analytica Chimica Acta,* 547, 2005, pp. 132-137.

Tsalev, D.L. (1999) - Hyphenated vapour generation atomic absorption spectrometric techniques, *Journal of Analytical Atomic Spectrometry,* 14, 1999, pp. 147-162.

Tsalev, D.L. (2000) – Vapor generation or electrothermal atomic absorption spectrometry? – Both!, *Spectrochimica Acta Part B*, 55, 2000, pp. 917-933.

Vale, M.G.R., Oleszczuk, N., dos Santos, W.N.L. (2006) – Current status of direct solid sampling for electrothermal atomic absorption spectrometry. A critical review of the development between 1995 and 2005, *Applied Spectroscopy Reviews*, 41, 2006, pp. 377-400.

Vereda Alonso, E., García de Torres, A., Cano Pavón, J.M. (2001) - Flow injection on-line electrothermal atomic absorption spectrometry. Review, *Talanta*, 55, 2001, pp. 219-232.

Walsh, A. (1991) – The development of atomic absorption methods of elemental analysis 1952-1962, *Analytical Chemistry*, 64, 1991, pp. 933A-941A.

Welz, A., Becker-Ross, H., Florek, S., Heitmann, U., Vale, M.G.R. (2003) - High-resolution continuum-source atomic absorption spectrometry – What can we expect?, *Journal of the Brazilian Chemical Society*, 14, 2003, pp. 220-229.

Villa-Lojo, M.C., Alonso-Rodríguez, E., López-Mahía, P., Muniategui-Lorenzo, S., Prada-Rodríguez, D. (2002) – Coupled high performance liquid chromatography – microwave digestión – hydride generation – atomic absorption spectrometry for inorganic and organic arsenic speciation in fish tissue, *Talanta*, 57, 2002, pp. 741-750.

Wolf, W.R, Stewart, K.K. (1979) – Automated multiple flow injection analysis for flame atomic absorption spectrometry, *Analytical Chemistry*, 51, 1979, pp. 1201-1205.

Yebra, M.C. (2000) - Continuous automatic determinations of organic compounds by flow injection – atomic absorption spectrometry, *Trends in Analytical Chemistry*, 19, 2000, pp. 629-641.

Yoza, N., Aoyagi, Y., Obashi, S., Tateda, A. (1979) - Flow injection system for atomic absorption spectrometry, *Analytica Chimica Acta*, 111, 1979, pp. 163-167.

Zybin, A., Koch, J., Butcher, D.J., Niemax, K. (2004) - Element-selective detection in liquid and gas chromatography by diode laser absorption spectrometry, *Journal of Chromatography A*, 1050, 2004, pp. 35-44.

Zybin, A., Koch, J., Wizemann, H.D., Franzke, J., Niemax, K. (2005) - Diode laser atomic absorption spectrometry. Review. *Spectrochimica Acta Part B*, 60, 2005, pp. 1-11.

8.5. BIBLIOGRAPHY

8.5.1. Books

Broekaert J.A.C. (2005) - Analytical Atomic Spectrometry with Flames and Plasmas. 2nd Completely revised and enlarged edition, Wiley-VCH, Weinheim, Germany, 2005, ISBN: 3-527-31282-X, 420 pp.

Butcher D.J., Sneddon J. (1998) – *A Practical Guide to Graphite Furnace Atomic Absorption Spectrometry,* John Wiley & Sons Ltd., Chichester, England, 1998, ISBN:0-471-12553-9, 272 pp.

Cullen M. (Editor) (2003) - *Atomic Spectroscopy in Elemental Analysis,* Blackwell Publishing, Oxford, England, 2003, ISBN: 1-84127-333-3, 328 pp.

Dean J.R., Ando D.J. (Editor) (1997) - *Atomic Absorption and Plasma Spectroscopy, 2nd Edition,* John Wiley & Sons Ltd., Chichester, England, 1997, ISBN: 0-471-97255-X, 228 pp.

Dedina J., Tsalev D.L. (1995) - *Hydride Generation Atomic Absorption Spectrometry,* John Wiley & Sons Ltd., Chichester, England, 1995, ISBN: 0-471-95364-4, 544 pp.

Ebdon L., Evans E.H., Fisher A.S., Hill S.J. (1998) - *An Introduction to Analytical Atomic Spectrometry,* John Wiley & Sons Ltd., Chichester, England, 1998, ISBN: 0-471-97417-X, 206 pp.

Fang Z. (1995) – *Flow Injection Atomic Absorption Spectrometry,* John Wiley & Sons Ltd., Chichester, England, 1995, ISBN: 0-471-95331-8, 306 pp.

Haswell S.J. (1991) - *Atomic Absorption Spectrometry,* Elsevier Science, Amsterdam, The Netherlands, 1991, ISBN: 0-444-88217-0, 530 pp.

Inczédy J., Lengyel T., Ure A.M. (1998) - *International Union of Pure and Applied Chemistry. Compendium of Analytical Nomenclature. Definitive rules 1997,* Blackwell Science, Oxford, England, 1998, ISBN: 0-86542-615-5, 964 pp.

Jackson K.W. (Editor) (1999) - *Electrothermal Atomization for Analytical Atomic Spectrometry,* John Wiley & Sons Ltd., Chichester, England, 1999, ISBN: 0-471-97425-0, 484 pp.

Jenniss S.W., Katz S.A., Lynch R.W. (1997) - *Applications of Atomic Spectrometry to Regulatory Compliance Monitoring, 2nd Edition,* Wiley-VCH, Weinheim, Germany, 1997, ISBN: 0-471-19039-X, 248 pp.

Sanz-Medel A. (Editor) (1999) - *Flow Analysis with Atomic Spectrometric Detectors,* Elsevier Science B.V., Amsterdam, The Netherlands, ISBN: 0-444-82391-3, 467 pp.

Tsalev D.L. (1995) - *Atomic Absorption Spectrometry in Occupational and Environmental Health Practice,* CRC Press, Boca Raton, Florida, USA, 1995, ISBN 0-8493-4999-0, 349 pp.

Welz B., Becker-Ross H., Florek S., Heitmann U. (2005) - *High-Resolution Continuum Source AAS: The Better Way to Do Atomic Absorption Spectrometry,* Wiley-VCH, Weinheim, Germany, 2005, ISBN: 3-527-30736-2, 296 pp.

Welz B., Sperling M. (1998) - *Atomic Absorption Spectrometry, 3rd Completely Revised Edition,* Wiley-VCH, Weinheim, Germany, 1998, ISBN: 3-527-28571-7, 965 pp.

8.5.2. Journals

The list below (ordered alphabetically) lists the journals probably known most worldwide (and with high impact factors as well) that are used by scientists to publish both, instrumental developments and new analytical applications with AAS techniques.

Analytica Chimica Acta

Analytical and Bioanalytical Chemistry

Analytical Chemistry

Analytical Letters

Analytical Sciences

Applied Spectroscopy

Journal of Analytical Atomic Spectrometry

Mikrochimica Acta

Spectrochimica Acta Part B

Talanta

The Analyst

INDEX

www.ingramcontent.com/pod-product-compliance
Ingram Content Group UK Ltd.
Pitfield, Milton Keynes, MK11 3LW, UK
UKHW021312070726
13610UKWH00001B/3